Thomas W. Wieting

2009

The Concept of a Riemann Surface

THIRD EDITION

Hermann Weyl

Translated from the German by
Gerald R. MacLane

Dover Publications, Inc.
Mineola, New York

Bibliographical Note

This Dover edition, first published in 2009, is an unabridged republication of the work originally published by Addison-Wesley Publishing Company, Inc., Reading, Massachusetts, in 1955.

Library of Congress Cataloging-in-Publication Data

Weyl, Hermann, 1885–1955.
[Ider der Riemannschen Fläche German]
The concept of a Riemann surface / Hermann Weyl. — Dover ed.
p. cm.
Originally published: 3rd ed. Reading, Mass. : Addison-Wesley, 1955.
Includes index.
ISBN-13: 978-0-486-47004-7
ISBN-10: 0-486-47004-0
1. Riemann surfaces. I. Title.

QA333.W413 2009
515'.93—dc22

2008051084

Manufactured in the United States of America
Dover Publications, Inc., 31 East 2nd Street, Mineola, N.Y. 11501

Dedicated to my sons

Joachim and Michael

PREFACE

This book was first published in 1913. It contained the essentials of some lectures which I gave at the University of Göttingen in the winter semester of 1911–12. The purpose of the book was to develop the basic ideas of Riemann's theory of algebraic functions and their integrals and also to treat the requisite ideas and theorems of analysis situs in a fashion satisfying modern demands of rigor. This had not been done before. For example, the concept of a curve is never clarified in the classic book of Hensel and Landsberg, *Theorie der algebraischen Funktionen einer Variablen und ihre Anwendung auf algebraische Kurven und Abelsche Integrale* (Leipzig 1902, Teubner). Three events had a decisive influence on the form of my book: the fundamental papers of Brouwer on topology, commencing in 1909; the recent proofs by my Göttingen colleague P. Koebe of the fundamental uniformization theorems; and Hilbert's establishment of the foundation on which Riemann had built his structure and which was now available for uniformization theory, the Dirichlet principle. The book was dedicated to Felix Klein "in Dankbarkeit und Verehrung." Klein had been the first to develop the freer conception of a Riemann surface, in which the surface is no longer a covering of the complex plane; thereby he endowed Riemann's basic ideas with their full power. It was my fortune to discuss this thoroughly with Klein in divers conversations. I shared his conviction that Riemann surfaces are not merely a device for visualizing the many-valuedness of analytic functions, but rather an indispensable essential component of the theory; not a supplement, more or less artificially distilled from the functions, but their native land, the only soil in which the functions grow and thrive. Even more than the text, the enthusiastic preface betrayed the youth of the author.

In 1923 Teubner published an anastatic reprint to which were added a page of corrections and additions and an appendix, "A rigorous foundation of the theory of characteristics on two-sided surfaces." This second edition has been distributed since 1947 in an American reprint, authorized by the Attorney General, by the Chelsea Publishing Co.

In the more than forty years which have passed since the appearance of this book the face of mathematics has changed noticeably. Above all, the young shoot "analysis situs" has become the tree topology, affording shade to large parts of our science. When German mathematicians and the publishing house of Teubner approached me with the invitation to prepare a new edition, since requests for the book continued, it at first seemed appropriate to treat the book more or less as an historical document and send it into the world again unchanged except for a few minor improvements. But as I attempted to merge the appendix with the main text, I became ever more conscious of the deficiencies of both the appendix and the text. The way in which I then undertook to rework the first topological half of the book more thoroughly, emphasizing the combinatorial aspect even more than formerly, may be seen from my paper in Die Zeitschrift für angewandte Mathematik und Physik (**4**, 1953, 471–492): *Über die kombinatorische und kontinuumsmässige Definition der Überschneidungszahl zweier geschlossener Kurven auf einer Fläche*. However, there occurred to me during the course of the work the idea, worked out in the end of that paper, of defining the intersection number by "topologizing" the construction I used in 1913 to define the Abelian integrals of the first kind. This resulted in a clearer structure for the second function-theoretic part of the book; in this development the following important fact no longer had to be suppressed, namely that the imaginary parts of the integrals of the first kind associated with the paths of a basis are the coefficients of a positive quadratic form. Also I found myself in agreement with the recent trend of topology, to replace the dissection of a manifold by a covering by overlapping neighborhoods. To be sure, it then turned out to be natural to subsume Riemann surfaces under differentiable surfaces, rather than the most general topological surfaces. The separation into real and imaginary parts, used systematically in the first edition, which obviates reference to a canonical dissection in Riemann's sense, remained fundamental. The effectiveness of this step is shown most clearly by the generalization to higher dimensions which was completed in the interim: the real harmonic, not the complex holomorphic, linear differential forms have established themselves as the starting point of the theory, and the great paper of Kunihiko Kodaira, *Harmonic fields in Riemannian manifolds (generalized potential theory)*, Ann. of Math. **50** (1949) 587–665, specifically invokes the prototype in my old book.

I remark further that in using a covering of the manifold by neighborhoods I find it necessary neither to normalize the neighborhoods (e.g., by the condition of convexity) nor to replace the covering by always finer ones. Also, the cycles, that is, the continuous closed curves, and the integrals along them, take their place in the most natural fashion alongside cocycles, which are so much more accommodating in general topology.

Only subsequently did I observe how much closer my new presentation has come to Claude Chevalley's treatment of Riemann surfaces in chapter

VII of his *Introduction to the theory of algebraic functions of one variable* (Math. Surveys VI, Am. Math. Soc., New York 1951). With Chevalley I must accept the reproach of André Weil that the route we chose ("without triangulation") has barred, or at least made more difficult, the way to certain classical results; for example, the structure of the fundamental group. (See Weil's review of Chevalley's book, Bull. Am. Math. Soc. **57** (1951) 384–398; in particular p. 390.) Since these results are beyond the scope of my book I felt that I should accept these limitations for the sake of the advantages of the method. Who says to us that we have already reached the end of the methodical development of topology?

As for references, I have preserved the old-fashioned aspect of the book. I have retained the citations to the literature of the 19th century; the younger generation is only too inclined to forget the connection of the new with the old! The references have been completed by citations of the newer literature, but I have not troubled to prepare a bibliography which is complete in any sense. The reader will find a welcome supplement to the algebraic aspects in the book of Chevalley noted above. Related subjects are treated in: R. Courant, *Dirichlet's principle, conformal mapping, and minimal surfaces,* New York 1950, and Rolf Nevanlinna, *Uniformisierung,* Berlin 1953; these works contain extensive references to the literature. There is as yet no comprehensive source for the splendid developments in "several variables" which were introduced in the book of W. V. D. Hodge, *The theory and applications of harmonic integrals,* Cambridge 1941, and in which the paper of Kodaira cited above is such a landmark. I have restricted myself to the case of *one* complex dimension, and throughout I use either the real or complex numbers as the coefficient domain, in the spirit of the "transcendental" Riemannian point of view which I adopt.

Influenced by Kodaira's work, I have hesitated a moment as to whether I should not replace the Dirichlet principle by the essentially equivalent "method of orthogonal projection" which is treated in a paper of mine in the Duke Math. Jour. **7** (1940), 411–444. But for reasons the explication of which would lead too far afield here, I have stuck to the old approach.

I fear that in the preparation of the new edition I have followed my own ideas too much and have paid too little attention to other ideas, especially those in the potent literature of topology. May I not be judged too harshly! For technical assistance in preparation of the manuscript I wish to thank most sincerely Frau Natascha Artin, in whom the mathematicians at Princeton and at New York University always have a friend ready to help. My thanks are also due to the publishing house of B. G. Teubner in Stuttgart who, though still in the middle of recreating their business, expended the same care on this new edition as the old firm granted to the first edition before the first world war.

HERMANN WEYL

Princeton, New Jersey
January 1955

Translator's note: Dr. Weyl's (1885–1955) academic affiliations included: Ph. D., Göttingen, 1908; hon. Ph. D., Oslo, 1929; hon. Dr. Ing., Stuttgart, 1929; hon. Dr. Sc., Penn., 1940; hon. Dr. Math., Eid. Tech. Hochs. Zürich, 1945; Privat-Dozent, Göttingen, 1910–1913; Prof. of Math., Eid. Tech. Hochs. Zürich, 1913–1930; Prof. of Math., Göttingen, 1930–1933; Professor, Institute for Advanced Study, 1933–1951; Professor Emeritus, 1951–1955. Professor, Princeton, 1928–1929. Lobatschefski Prize, Kazan, 1925.

CONTENTS

I. CONCEPT AND TOPOLOGY OF RIEMANN SURFACES

§ 1. Weierstrass' concept of an analytic function

Let z be a complex variable and a a fixed complex number. With Weierstrass we say that any power series

$$(1.1) \qquad \mathfrak{P}(z-a) = A_0 + A_1(z-a) + A_2(z-a)^2 + \cdots,$$

with positive radius of convergence, is a *function element with center a*. The coefficients A_0, A_1, $A_2 \ldots$ are arbitrary complex numbers. The region of convergence of such a power series consists either of the whole z-plane or of a disc $|z-a| < r \, (r > 0)$, the "*convergence disc,*" and a subset[1] of the periphery $|z-a| = r$ of that disc.

In its convergence disc (which may be the whole plane regarded as a disc of radius $r = \infty$), such a function element represents a regular analytic function in the sense of Cauchy. Conversely, it is known from elementary function theory that a uniform regular analytic function may be expanded in a convergent power series (1.1) in any neighborhood $|z-a| < r$ which is contained in the domain of regularity of the function. A power series then serves to represent the function only in a circular part of its domain.

If one starts with a power series which defines the function only in the convergence disc of the series (1.1), then the goal must be to define the function in larger domains of the z-plane without losing the analytic character of the function. The method for this is Weierstrass' principle of analytic continuation.[2] It turns out that the plan to conquer a largest possible domain

---

[1]) The term "subset" (of a set) is to be understood to include both the whole set and the void set which contains no element.

[2]) See the paper of Weierstrass, *Definition analytischer Funktionen einer Veränderlichen vermittelst algebraischer Differentialgleichungen,* which was written in 1842 but was first published in his *Mathematische Werke,* **1** (1894) 83–84. See also the first pages in Riemann's *Theorie der Abelschen Funktionen* (1857), [*Werke,* 2nd edition, pp. 88–89]. See also the article by Osgood in *Encyklopädie der math. Wiss.* II B 1, no. 13.

of the z-plane, for the function to be defined, is possible in only one way. But the uniformity (single-valuedness) of the function is usually lost in the process of analytic continuation. This is not to be regarded as a defect; rather it is a great merit that in this fashion also the many-valued analytic functions become amenable to an exact treatment.

If b is a value of z in the convergence disc $|z - a| < r$, then, as one knows, a rearrangement of the series (1.1) in powers of $z - b$ gives a new power series,

$$\mathfrak{Q}(z - b) = B_0 + B_1(z - b) + B_2(z - b)^2 + \cdots, \tag{1.2}$$

which converges at least in the disc $|z - b| < r - |b - a|$; its convergence disc may have a radius greater than $r - |b - a|$. Since (1.1) and (1.2) take the same values at each point common to their two convergence discs, (1.2) provides an extension of the definition of our analytic function beyond the original domain. We shall say that (1.2) is an *immediate analytic continuation* of (1.1). The general process of (mediate) *analytic continuation* consists in applying immediate analytic continuation not just once, but an arbitrary finite number of times in a sequence – which is approximately analogous to the fact of projective geometry that the general projective transformation may be obtained by a sequence of an arbitrary number of immediate projective, i.e., perspective, transformations.

Analytic continuation may be undertaken *along a given curve* $\mathfrak{c}$. This means the following. Let a curve starting at $z = a$ be given, i.e., to each real λ in the interval $0 \leq \lambda \leq 1$ there corresponds in continuous fashion a point z_λ of the complex z-plane; in particular, to $\lambda = 0$ there corresponds the point $z_0 = a$, and to $\lambda = 1$ a point $z = c$. Also, to every value of the parameter λ let there correspond a function element $\mathfrak{P}_\lambda$ with center z_λ; let $\mathfrak{P}_0$ be the given function element (1.1) and also let the following condition hold: if λ_0 is any λ-value, then there exists a positive number ε such the points z_λ which correspond to values λ in the interval

$$\max(0, \lambda_0 - \varepsilon) \leq \lambda \leq \min(1, \lambda_0 + \varepsilon)$$

all lie in the convergence disc of $\mathfrak{P}_{\lambda_0}$ and the associated function elements $\mathfrak{P}_\lambda$ are all immediate analytic continuations of $\mathfrak{P}_{\lambda_0}$. Then we say that we have continued the given element (1.1) analytically along $\mathfrak{c}$, and we call $\mathfrak{P}_1$ the function element with center c which results from the analytic continuation of (1.1) along $\mathfrak{c}$. Conversely, the analytic continuation of $\mathfrak{P}_1$ backwards along $\mathfrak{c}$ results in $\mathfrak{P}_0$.

The analytic continuation along a given curve $\mathfrak{c}$ *is unique if it is possible at all.* The proof is based on a theorem which belongs to the foundations of analysis and which we will discuss in a more systematic context in § 4. It is the following: if to each λ in the unit interval $0 \leq \lambda \leq 1$ there corresponds not only the point z_λ, in continuous fashion, but also a disc $|z - z_\lambda| < r_\lambda$ about z_λ with positive radius r_λ, then the curve defined by z_λ can be partitioned into a finite number of arcs, $\lambda_i \leq \lambda \leq \lambda_{i+1}$ $(0 = \lambda_0 < \lambda_1 < \cdots < \lambda_n = 1)$, and a point μ_i $(\lambda_i < \mu_i < \lambda_{i+1})$ may be chosen on each arc, so that the ith arc is contained in the disc about $z_{\mu_i} = z_i$ of radius $r_{\mu_i} = r_i$ $(i = 0, 1, \ldots, n-1)$. If now $\mathfrak{P}_\lambda$ and $\mathfrak{P}^*_\lambda$ are two analytic continuations along the curve z_λ, choose the positive number r_λ so that $\mathfrak{P}_\lambda$ and $\mathfrak{P}^*_\lambda$ both converge in the disc about z_λ with radius $3r_\lambda$. Then not only do $\mathfrak{P}_{\mu_i}$ and $\mathfrak{P}^*_{\mu_i}$ converge in the disc of radius $3r_i$ about z_{μ_i}, but $\mathfrak{P}_{\lambda_i}$ and $\mathfrak{P}^*_{\lambda_i}$ converge in the disc of radius $2r_i$ about z_{λ_i}; the complete arc z_λ for $\lambda_i \leq \lambda \leq \lambda_{i+1}$ is contained in this disc, and all the associated $\mathfrak{P}_\lambda$ and $\mathfrak{P}^*_\lambda$ arise by immediate analytic continuation of $\mathfrak{P}_{\lambda_i}$ and $\mathfrak{P}^*_{\lambda_i}$. Thus the equation $\mathfrak{P}_{\lambda_i} = \mathfrak{P}^*_{\lambda_i}$ entails $\mathfrak{P}_\lambda = \mathfrak{P}^*_\lambda$ for $\lambda_i \leq \lambda \leq \lambda_{i+1}$. By proceeding from division point to division point, one obtains from $\mathfrak{P}_0 = \mathfrak{P}^*_0$ the equation $\mathfrak{P}_\lambda = \mathfrak{P}^*_\lambda$ for the whole unit interval and in particular for $\lambda = 1$; thus the last element is uniquely determined by the initial element and the curve along which the continuation takes place.

These considerations also show that if continuation along a given curve $\mathfrak{c}$ is possible, then one can get from the initial element to the last element by a finite number of applications of immediate analytic continuation. If the continuation of the initial element along $\mathfrak{c}$ is impossible, then there exists a definite point on the curve, the "critical point," at which the process finds its necessary end. More precisely: there exists a threshold $\lambda = \Lambda_0$ of the following sort. If $\lambda_0 < \Lambda_0$, then analytic continuation along the subcurve $z = z_\lambda$ $(0 \leq \lambda \leq \lambda_0)$ can be carried out; but not if $\lambda_0 = \Lambda_0$. Naturally $\Lambda_0 = 1$ is possible; i.e., the endpoint of $\mathfrak{c}$ is the critical point; on the other hand, we always have $\Lambda_0 > 0$.

Still another theorem on analytic continuation is important. If

$$z = z_1(\lambda), \qquad z = z_2(\lambda)$$

are two curves, from the same point $a\{= z_1(0) = z_2(0)\}$ to the same endpoint c, which remain sufficiently close together, then if the analytic continuation is possible along the first curve, it is possible along the second and yields the same end element. The condition that the curves remain sufficiently close together is: there exists a positive number δ such that if $|z_1(\lambda) - z_2(\lambda)| < \delta$ for all λ, then the conclusion of the theorem is valid. The proof follows immediately from the fact that one can obtain the end element from the initial element by a finite chain of immediate analytic continuations.

Now we are in a position to state the general Weierstrass definition of an analytic function as follows: *an* **analytic function** *is the totality G of all those function elements which can arise from a given function element by analytic continuation.*

Every function element of G may be obtained from any other by analytic continuation.

It can be shown that if the two analytic functions G_1 and G_2 have a single function element in common, then they are identical; i.e., every element of G_1 is an element of G_2 and conversely.

If

$$\mathfrak{P}(z-a) = A_0 + A_1(z-a) + A_2(z-a)^2 + \cdots$$

is an element of G, then A_0 is called a *value* of the analytic function G at the point $z = a$.

Certainly, at first glance, there is something artificial about Weierstrass' concept of a many-valued analytic function as a *collection of function elements.* When one talks of $\sqrt{z}$ or $\log z$, one hardly envisages the totality of power series which represent pieces of these many-valued functions. Nevertheless, Weierstrass' definition, whose simplicity and precision cannot be denied, has the advantage of being a solid starting point for analytic function theory. By gradual reworking of Weierstrass' formulation we will arrive at Riemann's formulation, in which the independent variable z as well as the dependent variable u, which up to now is represented by a totality G of function elements, appear as uniform analytic functions of a parameter; a parameter, to be sure, which in general takes values not in the complex plane but on a certain two-dimensional manifold, the so-called Riemann surface.

But first we must extend, with Weierstrass, the concept of analytic function to that of analytic form.

§ 2. The concept of an analytic form

The concept of the analytic form arises from that of an analytic function when one considers not merely the points where the function is regular, as has been the case up to now, but adds those points at which it has a branch point of finite order or a pole (or both at once). If we suitably generalize the previous concept of function element, then we obtain the precise formulation of the concept of an analytic form.[3]

---

[3]) See Weierstrass, *Vorlesungen über die Theorie der Abelschen Transzendenten* (edited by G. Hettner and I. Knoblauch), *Werke,* **4**, 16–19.

With the aid of a complex parameter t we can represent the function element (1.1): $u = \mathfrak{P}(z - a)$, as follows:

$$z = a + t, \qquad u = \mathfrak{P}(t) = A_0 + A_1 t + A_2 t^2 + \cdots .$$

If we abandon the distinguished role played by z and also allow a finite number of negative powers of t, we obtain the more general formulation. Let

$$z = P(t), \qquad u = Q(t)$$

be any two series in integral powers of t which contain only a finite number of terms with negative powers of t and which are such that in some neighborhood $|t| < r$ (r a positive constant) of the origin: (1) both series converge, and (2) no two different values of t in this neighborhood give the same pair of values (z, u). Then we say that this pair of power series *defines a function element*. We add the condition that $P(t)$ is not a mere constant.

It is not our intention that the pair of power series $P(t)$, $Q(t)$ is understood to be the "function element"; rather we regard the two series only as a *representation* of the intended function element, which has infinitely many other representations with equal claims. Concerning the transformation of one representation to another, we make the following agreement.

If one substitutes the power series

$$t(\tau) = c_1 \tau + c_2 \tau^2 + \cdots$$

for the parameter t in both $P(t)$ and $Q(t)$, then $P(t)$ turns into a power series $\Pi(\tau)$, $Q(t)$ into $\mathsf{K}(\tau)$. We assume that $t(\tau)$ converges in some neighborhood of $\tau = 0$ and that the first coefficient $c_1 \neq 0$; then there is a positive constant ρ such that in $|\tau| < \rho$, $t(\tau)$ (1) converges and has modulus $< r$, and (2) assumes different values at any two distinct points τ. Then, in this neighborhood $|\tau| < \rho$, Π and K converge, and for any two distinct points τ_1, τ_2 of this neighborhood, not both equations $\Pi(\tau_1) = \Pi(\tau_2)$, $\mathsf{K}(\tau_1) = \mathsf{K}(\tau_2)$ hold. We say the pair Π, K is *equivalent* to the original pair P, Q, no matter what the coefficients $c_1, c_2, \ldots$ in the series $t(\tau)$ are, provided only that $t(\tau)$ converges and $c_1 \neq 0$. The last assumption has the consequence that conversely $P(t)$, $Q(t)$ may be obtained from $\Pi(\tau)$, $\mathsf{K}(\tau)$ by substituting for τ a certain power series in t:

$$\tau = \gamma_1 t + \gamma_2 t^2 + \cdots \qquad \left(\gamma_1 = \frac{1}{c_1} \neq 0\right).$$

The relation of equivalence is thus symmetric. Also it is obvious that any pair of power series is equivalent to itself, and that if two pairs of power series are equivalent to a third, then they are equivalent to each other. These facts justify us in regarding equivalent pairs of power series as representations of *the same*, and nonequivalent pairs as representations of *different*, function elements. Or, to restate it: *two pairs of power series, each representing a function element, define the same function element if and only if they are equivalent.*

We depend here on a method of definition which one must use frequently in mathematics and which has its psychological roots in our minds' capability for *abstraction*. This kind of definition rests on the following general principle. If between the objects of any domain of operation there is specified a relation $\sim$ *which has the character of equivalence*; i.e., a relation satisfying the laws

$$(1)\ a \sim a, \qquad (2) \text{ from } a \sim b \text{ follows } b \sim a,$$
$$(3) \text{ from } a \sim c, \quad b \sim c \text{ follows } a \sim b;$$

then it is possible to regard each object a of that original domain of operation as *a representative of an object* α such that two objects a, b are representatives of the same object α if and only if they are equivalent in the sense of the relation $\sim$. Precisely this principle is always to be used when we are interested only in those properties of the objects $a, b, \ldots$ which are invariant under the relation $\sim$. Its application has the advantage that a cumbersome terminology is replaced by a shorter one which suits the center of interest of the investigation in that it automatically *strips* the objects of what is *inessential* relative to this center. I mention here two examples of such "*Definition by Abstraction.*"

(1) One says that two parallel lines have the same *direction*; two nonparallel lines have different directions. The original objects (a) are the lines; the relation with the character of equivalence is parallelism. One wishes to associate with each line a "something," its "direction," so that parallelism of lines corresponds to the identity of the associated "directions."

(2) A "motion" (of a point) is specified if the position of the moving point p is given at each instant λ of a certain time interval $\lambda_0 \leq \lambda \leq \lambda_1 : p = p(\lambda)$. If one has two such motions, $p = p(\lambda)$, $q = q(\mu)$, then one says these motions travel the same "*path*" if and only if λ, the time parameter of the first motion, can be expressed as a continuous monotone increasing function of the time parameter μ of the second motion, $\lambda = \lambda(\mu)$, such that thereby the first motion becomes the second: $p(\lambda(\mu)) \equiv q(\mu)$. Here it is the concept of "path" which is to be defined.[4]

---

[4]) This concept is something more than that of the *point set* which consists of all points passed in the motion. We are concerned with the same distinction as, in the case of a pedestrian, that between the path traced (which, as long as he walks, is in statu nascendi) and the path (long since existing) *on which* he walks.

We return from this digression to our extended concept of a function element. From among all the equivalent representations of one and the same function elemcnt we shall attempt to single out one, the *normal representation,* which is as simple as possible. We consider several cases. If $P(t)$ contains no negative powers of t,

$$z = P(t) = a + a_1 t + a_2 t^2 + \cdots,$$

and if $a_1 \neq 0$, then we can introduce $z - a = \tau$ as a new parameter to obtain

$$z = a + \tau, \qquad u = K(\tau) \tag{2.1}$$

as a new representation of the same function element. If Q also contains no negative powers, then the same is true of K, and we have a function element of the type considered in § 1, which we shall now, to distinguish it, designate as a *regular function element.* (2.1) is the desired normal representation. A regular function element has only one normal representation, and therefore two regular function elements are certainly different if their normal representations differ. By virtue of this fact, and only now, are we really justified in calling our present concept of function element a generalization of the concept in § 1.

If we assume that the expansion of z contains no negative powers of t but, more generally,

$$z = a + a_\mu t^\mu + a_{\mu+1} t^{\mu+1} + \cdots \qquad (a_\mu \neq 0),$$

that is, if, aside from the constant term, a_μ $(\mu \geq 1)$ is the first nonzero coefficient, then one can substitute for t a power series

$$t = c_1 \tau + c_2 \tau^2 + \cdots \qquad (c_1 \neq 0)$$

so that

$$z = a + \tau^\mu,$$

and we obtain a representation of the form

$$z = a + \tau^\mu, \qquad u = K(\tau). \tag{2.1*}$$

In fact it is known that if $\sqrt[\mu]{a_\mu}$ is a given one of the μ roots, then there exists a unique power series $\gamma_1 + \gamma_2 t + \gamma_3 t^2 + \cdots$, with the leading term $\gamma_1 = \sqrt[\mu]{a_\mu}$, whose μth power $= a_\mu + a_{\mu+1} t + \cdots$. By solving

$$\tau = \gamma_1 t + \gamma_2 t^2 + \cdots$$

for t, we obtain the desired result. But by using the μ different roots $\sqrt[\mu]{a_\mu}$, one obtains μ distinct representations (2.1*); they may all be obtained from one of them by replacing τ by $\tau\zeta$, where ζ is an arbitrary μth root of unity. A function element given by

$$(2.2) \qquad z = a' + \tau^{\mu'}, \qquad u = \mathsf{K}'(\tau) \qquad (\mu' \text{ a positive integer})$$

is the same as that of (2.1*) if and only if $a' = a$, $\mu' = \mu$, and (2.2) is the same, coefficient by coefficient, as one of the μ normal representations arising from (2.1*) by the substitution $\tau|\tau\zeta$. In particular, the integer μ is characteristic of the given function element and independent of the particular representation. We say that the element is *branched, with order* $\mu - 1$; in the case $\mu = 1$, which we considered above, the element is called *unbranched.*

If the expansion of z contains negative powers of t, let $t^{-\nu}$ be the lowest one:

$$z = a_{-\nu}t^{-\nu} + a_{-\nu+1}t^{-\nu+1} + \cdots \qquad (a_{-\nu} \neq 0).$$

One can replace t by a convergent power series in τ, $t = c_1\tau + \cdots$ $(c_1 \neq 0)$ so that

$$(2.1^{**}) \qquad z = \tau^{-\nu}, \qquad u = \mathsf{K}(\tau).$$

This is the normal representation in this case; if $\nu > 1$, the representation is not uniquely determined by the element, rather there are ν distinct representations which may all be obtained from any one by replacing τ by $\tau\zeta$, where ζ is an arbitrary νth root of unity. Here also one speaks of a branching of order $\nu - 1$.

In the derivation of the normal representations (2.1), (2.1*), and (2.1**) we have, as in § 1, given the variable z preference over u. The irregular function elements have appeared alongside the regular ones and the branched elements alongside the unbranched ones. Now by extending the concepts of immediate and mediate analytic continuation to arbitrary (including irregular) function elements, we arrive without more ado at the definition of an analytic form.

Let $\mathfrak{e}$ be a function element and let

$$(2.3) \qquad z = P(t), \qquad u = Q(t)$$

be any representation of $\mathfrak{e}$ *valid* in the disc $|t| < r$ $(r > 0)$ [i.e., $P(t)$ and $Q(t)$ converge in this disc, and at most one of the equations $P(t_1) = P(t_2)$, $Q(t_1) = Q(t_2)$ holds for $t_1 \neq t_2$, $|t_1| < r$, $|t_2| < r$]. For every value t_0,

$|t_0| < r$, we can rearrange the series $P(t)$, $Q(t)$ in powers of $t' = t - t_0$ and obtain a new pair of power series $P'(t')$, $Q'(t')$ and thus a new function element $\mathfrak{e}_{t_0}$. We say that all the elements $\mathfrak{e}_{t_0}$ for $|t_0| < r$ constitute an *analytic neighborhood* of the original element $\mathfrak{e}$ $(= \mathfrak{e}_0)$; this terminology will be used no matter which possible parameter t is used to represent $\mathfrak{e}$ and no matter which disc $|t| < r$ in which the representation is valid is chosen. The analytic neighborhood described above, determined by the representation (2.3) together with the inequality $|t| < r$, will be called, whenever we need a short label, and with reference to the parameter of representation, a *t-neighborhood.*

If one has two different representations of a function element $\mathfrak{e}$,

$$z = P(t), \qquad u = Q(t),$$

and

$$z = \Pi(\tau), \qquad u = \mathrm{K}(\tau),$$

which are equivalent via the substitution

$$t = t(\tau) = c_1\tau + c_2\tau^2 + \cdots \qquad (c_1 \neq 0),$$

then the following important theorem is true.

Every t-neighborhood of $\mathfrak{e}$ *contains a τ-neighborhood of* $\mathfrak{e}$, *and conversely every τ-neighborhood contains a t-neighborhood.*

Proof. Let the t-neighborhood be specified by $|t| < r$. Choose a positive number ρ such that the power series $t(\tau)$ converges, remains in absolute value $< r$, and does not assume the same value twice, in $|\tau| < \rho$. This inequality $|\tau| < \rho$ determines a τ-neighborhood which, I claim, is contained in the original t-neighborhood. By rearrangement of $\Pi(\tau)$, $\mathrm{K}(\tau)$ in powers of $\tau' = \tau - \tau_0$ $(|\tau_0| < \rho)$, we obtain a representation $\Pi'(\tau')$, $\mathrm{K}'(\tau')$ of the element $\mathfrak{e}_{\tau_0}$. By rearrangement of $P(t)$, $Q(t)$ in powers of $t' = t - t_0$ $[t_0 = t(\tau_0), |t_0| < r]$, we get a representation of the element $(P'(t'), Q'(t')) = \mathfrak{e}_{t_0}$. Then $\mathfrak{e}_{\tau_0} = \mathfrak{e}_{t_0}$. For by rearranging $t(\tau)$ in powers of $\tau' = \tau - \tau_0$ we obtain

$$t'(\tau') = t(\tau) - t_0 = c_1'\tau' + c_2'\tau'^2 + \cdots, \tag{2.4}$$

and clearly

$$P'(t'(\tau')) = \Pi'(\tau'), Q'(t'(\tau')) = \mathrm{K}'(\tau');$$

for, to take the first equation, $\Pi'(\tau - \tau_0)$ and $P'(t(\tau) - t_0)$ are regular functions of τ in some neighborhood of the point τ_0 in the complex τ-plane which takes the same values as $\Pi(\tau)$. For the substitution (2.4), which carries the pair $P'(t')$, $Q'(t')$ into $\Pi'(\tau')$, $\mathrm{K}'(\tau')$, the first coefficient is $c_1' = [dt(\tau)/d\tau]_{\tau=\tau_0} \neq 0$; for otherwise the function $t(\tau)$ would take some value at least twice in each neighborhood of $\tau = \tau_0$. Thus we have proved the identity of $\mathfrak{e}_{\tau_0}$ and $\mathfrak{e}_{t_0}$.

The concept of analytic neighborhood introduced here corresponds, in a way adapted to further developments, to the concept of immediate analytic continuation used in § 1. With the introduction of analytic neighborhoods, it also becomes clear that the irregular elements are, compared to the regular, to be regarded as the exception; for if the analytic neighborhood of a given element is chosen sufficiently small, then it consists (with the possible exception of the element itself) exclusively of regular elements. To see this, one need only choose the disc $|t| < r$ determining the neighborhood of $\mathfrak{e} = (P(t), Q(t))$ small enough so that (except possibly at $t = 0$) $dP/dt \neq 0$ in $|t| < r$. For the normal representation of $\mathfrak{e}$, for which $P(t) = a + t^\mu$ or $P(t) = t^{-\nu}$, this is true without more ado in the complete neighborhood in which the representation is valid.

Thus to each point z_1 in the neighborhood $|z_1 - a| < r^\mu$ or $|z_1| > r^{-\nu}$ of the center a or ∞, except the center itself, there correspond μ or ν regular elements $P(z - z_1)$ *induced* by the given element. For later use it is important to note that the function element $\mathfrak{e}$ can be uniquely reconstructed from any one of its induced elements. For example, consider the case where $\mathfrak{e}$ is given by

$$z = a + t^\mu, \qquad u = \sum_{i=-n}^{\infty} A_i t^i \qquad (|t| < r)$$

and assume for the sake of simplicity that $a = 0$. Let z_1 be any z-value of absolute value ρ, where $0 < \rho < r^\mu$. Each of the μ regular elements $Q(z - z_1)$ induced by $\mathfrak{e}$ at z_1 converges in the disc $|z - z_1| < \delta$, where $\delta = \min(r^\mu - \rho, \rho)$. These elements induced at z_1 may all be obtained by analytic continuation of one of them along the circle $|z| = \rho$, and the exponent μ is characterized by the fact that one first returns to the starting element after traversing this circle μ times. At the same time this analytic continuation determines the function $Q(t)$ in the annulus $\rho - \delta < |z| < \rho + \delta$, and the coefficients A_h in the power series development may be computed from the integrals

$$A_h = \frac{1}{2\pi i}\int (Q(t)/t^{h+1})\, dt$$

taken over the circle $|t| = \sqrt[\mu]{\rho}$ in the t-plane. Another, possibly simpler, procedure is this: let $0 < |z_1| < (\frac{1}{2} r)^\mu$ and let $\sqrt[\mu]{z_1} = t_1$ be one of the μ determinations of this root. To this root corresponds a definite one of the μ regular elements $\mathfrak{Q}$ induced at z_1 by $\mathfrak{e}$. The development of $t^n \mathfrak{Q}$ in powers of

$t - t_1$ will have a convergence disc containing the point $t = 0$, and therefore will furnish $t^n Q(t)$ by a single direct analytic continuation. μ is characterized as the smallest exponent for which, for sufficiently large n, the convergence disc of $t^n Q$ contains the point 0.

An *analytic chain of function elements* is given if to each real value λ of the interval $0 \leq \lambda \leq 1$ there corresponds a function element $\mathfrak{e}(\lambda)$ satisfying the condition: if λ_0 is any value of the parameter λ, $\mathfrak{e}_0 = \mathfrak{e}(\lambda_0)$ the corresponding function element, and $\mathfrak{U}_0$ an arbitrary analytic neighborhood of $\mathfrak{e}_0$, then there exists a positive number ε such that $\mathfrak{e}(\lambda)$ belongs to $\mathfrak{U}_0$ for $|\lambda - \lambda_0| \leq \varepsilon$. The analytic chain *connects* the initial element $\mathfrak{e}(0)$ to the terminal element $\mathfrak{e}(1)$.

An **analytic form** *is a set G of function elements with the following properties:*

(1) *Any two function elements of G can be connected by an analytic chain whose elements all belong to G.*

(2) *It is impossible to extend G by adjoining more function elements so that the extended set still has the property* (1).

In particular, it follows from (2) that if the function element $\mathfrak{e}$ belongs to G, then G contains every analytic neighborhood of $\mathfrak{e}$.

We can make these specifications seem more natural by an analogy. Obviously the Weierstrass function element plays the same role in function theory as the *point* plays as the space element in geometry, say that of three-dimensional space. Thus we regard function element as analogous to point, and, as we speak of the three-dimensional space of points, so we speak of the "space (i.e., the totality) of function elements"; since the prescription of a function element depends on infinitely many continuous parameters, we must ascribe infinite dimension to this space. All concepts relating to *continuity* in three-dimensional space may be derived from one, that of the "*neighborhood* of a point" (for example, we may take as a neighborhood of a point the interior of any sphere centered at the point). In the space of function elements, the entities which we have named "analytic neighborhoods" will be the analog of the concept "neighborhood" in the space of points. Then an *analytic chain of function elements* corresponds exactly to a *continuous curve* in point-space. As a result of these prescriptions, the space of function elements possesses a structure essentially different from that of the familiar three-dimensional space; for while the space of points forms a single connected whole (any two of its points may be connected by a continuous curve), the infinite dimensional space of function elements falls apart into infinitely many (two-dimensional) "layers." Each such layer is a continuously (this means here analytically) connected whole, but the individual layers are not joined to each other anywhere. These "layers" are precisely the analytic forms. In Euclidean space any point set with the following properties is called a "*domain.*" (1) Each point of the set has a neighborhood completely contained in the set, and (2) any two points of the set may be joined by a continuous curve whose points are all in the set. Thus in the space of function elements an analytic form is a *domain which cannot be extended to a larger domain.* The desire, grounded in the analogy we have discussed, to regard the analytic form as a *two-dimensional manifold* leads us immediately into the heart of the Riemann–Klein approach. The familiar Riemann surfaces presented in elementary textbooks have exactly this significance: to represent each function element of the analytic form by a single point of the surface so that analytic chains of function elements appear as continuous curves on the Riemann surface.

But before we follow this line of thought further, we must investigate the relation between the two concepts "analytic function" and "analytic form."

§ 3. The relation between the concepts "analytic function" and "analytic form"

We remark first that if one has succeeded in continuing a regular function element analytically, in the sense discussed in § 1, along a given curve $z = z(\lambda)$ $(0 \leq \lambda \leq 1)$ [so that each value λ corresponds to a regular function element $\mathfrak{e}(\lambda)$ in which the development of z in powers of the representation parameter t begins with the constant term $z(\lambda)$], then one has obtained thereby an analytic chain of function elements in the sense of § 2. For, if λ_0 is any value of λ, and $\mathfrak{U}_0$ is an arbitrary analytic neighborhood of $\mathfrak{e}(\lambda_0) = \mathfrak{e}_0$, then, because $\mathfrak{e}_0$ is regular, one may choose the variable $z' = z - z(\lambda_0)$ as the representation parameter for $\mathfrak{e}_0$ and there exists a z'-neighborhood of $\mathfrak{e}_0$, given by $|z'| < r_0$, which is contained in $\mathfrak{U}_0$. If we pick an arc about $z(\lambda_0)$, given by the inequality $|\lambda - \lambda_0| \leq \varepsilon$ $(\varepsilon > 0)$, for which $|z(\lambda) - z(\lambda_0)| < r_0$, then the $\mathfrak{e}(\lambda)$ corresponding to points of this arc are obtained by rearranging $\mathfrak{e}_0$ in powers of $z - z(\lambda)$; hence the $\mathfrak{e}(\lambda)$ belong to the z'-neighborhood, $|z'| < r_0$, of $\mathfrak{e}_0$ and hence to the neighborhood $\mathfrak{U}_0$. The converse, that an analytic chain of regular function elements is obtained from the initial element by the process of analytic continuation described in § 1, needs no proof.

Any analytic chain $\mathfrak{e}(\lambda)$ $(0 \leq \lambda \leq 1)$ *of function elements contains only a finite number of irregular function elements (in particular, only a finite number of branched elements).*

This follows from the fact that each element has an analytic neighborhood consisting of regular elements, with the possible exception of the given element itself.

Suppose that the expansion of z in the representation of the function element $\mathfrak{e}(\lambda)$ begins with the constant term $z(\lambda)$ (for convenience we exclude the case where the expansion begins with negative powers of the representation parameter t). We shall consider the case where branched function elements occur among the $\mathfrak{e}(\lambda)$. Let $\mathfrak{e}(0)$ be regular, let λ_0 (< 1) be the smallest value of λ for which $\mathfrak{e}(\lambda)$ is irregular, and let $\mu - 1$ be the branch order of $\mathfrak{e}(\lambda_0)$. The elements $\mathfrak{e}(\lambda)$, $0 \leq \lambda < \lambda_0$, are obtained uniquely by analytic continuation along the given curve $z = z(\lambda)$. This situation changes at λ_0; the analytic continuation along the given curve may be extended, starting at $\mathfrak{e}(\lambda_0)$, in exactly μ different ways: the trunk branches into μ boughs along one of which the *given* analytic chain runs. Each of these boughs may branch repeatedly as one progresses, and the next branching of one bough may occur at a different spot than the next branching of another bough, and so on. Of course, a twig may end completely, when one reaches a critical point which is not merely a pole or branch point of finite order, before one arrives at $\lambda = 1$ by that route. From this description one can recognize how aptly the Riemannian term "branch point" matches the basic intuitive essence of the matter. To prove our claims the following simple considerations suffice. Let

$$z = z(\lambda_0) + t^\mu, \qquad u = Q(t)$$

be one of the μ normal representations of $\mathfrak{e}(\lambda_0)$ valid for $|t| < r$. Choose $\lambda_1 > \lambda_0$ so that $|z(\lambda) - z(\lambda_0)| < r^\mu$ is satisfied for $\lambda_0 \leq \lambda \leq \lambda_1$. Then there are μ curves in $|t| < r$ which start at $t = 0$ and which correspond to the one curve $z = z(\lambda)$ $(\lambda_0 \leq \lambda \leq \lambda_1)$ under the map $z = z(\lambda_0) + t^\mu$; these μ curves are congruent and may be obtained by rotating any one about the origin through an angle $2\pi/\mu$, $4\pi/\mu$, ... By continuing $Q(t)$ along each of these curves (immediate analytic continuation!), one obtains the μ possible continuations of $\mathfrak{e}(\lambda_0)$ along the given curve from $\lambda = \lambda_0$ to $\lambda = \lambda_1$. It is not claimed that each of these continuations can be extended to an analytic chain reaching the end $(\lambda = 1)$; in general that is false.

The fact that each analytic chain contains only a finite number of irregular elements makes it possible to *avoid* these irregular elements. Two function elements which can be connected by an analytic chain may be connected by an analytic chain, all of whose elements are regular, with the possible exception of the initial and terminal elements. To prove this we assume for the sake of simplicity that the analytic chain $\mathfrak{e}(\lambda)$ contains only *one* irregular element, $\mathfrak{e}(\lambda_0)$ $(0 < \lambda_0 < 1)$. Let

$$z = P(t), \qquad u = Q(t)$$

be a representation of $\mathfrak{e}(\lambda_0)$ valid for $|t| < r$; we choose r small enough so that the elements of the t-neighborhood $\mathfrak{U}_0$ of $\mathfrak{e}(\lambda_0)$ specified by $|t| < r$ are all regular, except for $\mathfrak{e}(\lambda_0)$ itself. Pick $\lambda_1 < \lambda_0$ and $\lambda_2 > \lambda_0$ so that all elements $\mathfrak{e}(\lambda)$ $(\lambda_1 \leq \lambda \leq \lambda_2)$ belong to $\mathfrak{U}_0$. Then $\mathfrak{e}(\lambda_1)$ is obtained by rearranging the given representation of $\mathfrak{e}(\lambda_0)$ in powers of $t - t_1$, where t_1 is a certain point of $|t| < r$; $\mathfrak{e}(\lambda_2)$ is obtained by rearranging in powers of $t - t_2$. The points t_1 and t_2 may be joined by a curve, in $|t| < r$, which does not pass through the origin. By associating with each point t_0 of this curve the function element obtained by rearranging $P(t)$ and $Q(t)$ in powers of $t - t_0$, we obtain an analytic chain, which connects $\mathfrak{e}(\lambda_1)$ and $\mathfrak{e}(\lambda_2)$ and which consists entirely of regular elements.

The following two theorems are a consequence of the facts we have proved.

(1) *The regular function elements of an analytic form constitute a single analytic function.*

(2) *Every analytic function consists of the regular function elements of an analytic form which is uniquely determined by the function.*

To these we add a further theorem.

(3) *The set of irregular function elements of an analytic form is countable.*

The proof depends on the following theorem of Poincaré and Volterra:[5] *an analytic form contains at most a countable infinity of regular function elements* $u = \mathfrak{P}(z - a)$ *with prescribed center* $z = a$. Each of these elements may be obtained from one of them, $\mathfrak{P}_1$, by regular continuation of $\mathfrak{P}_1$ along curves in the z-plane which start and end at a. For each such curve one can construct a polygonal path, with a finite number of segments, which stays close enough to the curve so that regular analytic continuation along the polygon is possible and leads to the same final function element as continuation along the curve. Also one can choose the polygon so that its vertices have rational coordinates relative to a; let these relative coordinates $z - a$ be

$$\frac{n_1' + in_1''}{n_1}, \quad \frac{n_2' + in_2''}{n_2}, \quad \ldots, \quad \frac{n_h' + in_h''}{n_h} \qquad (i = \sqrt{-1}),$$

[5]) Poincaré, Rendiconti del Circolo matematico di Palermo, **2** (1888) 197–200. Volterra, Atti della Reale Academia dei Lincei, ser. 4, IV₂, 355.

where n'_f, n''_f, and n_f are integers without any common factor and $n_f > 0$ ($f = 1, 2, \ldots, h$). With this polygon we associate the number

$$\sum_{f=1}^{h} |n'_f| + \sum_{f=1}^{h} |n''_f| + \sum_{f=1}^{h} n_f = N.$$

Among the polygons starting and ending at a, whose vertices have rational coordinates with respect to a, there are clearly only finitely many associated with the same number N. I arrange all these polygons in a sequence by choosing $N = 3, 4, 5, \ldots$ successively. Each of these polygons determines either one or no function element with center a, according as the regular continuation of $\mathfrak{P}_1$ along the polygon exists or not. Certainly one obtains in this fashion all of the regular function elements which belong to the analytic form and have center a; thus their countability is proved.

Instead of the z-plane I can use the z-sphere, the stereographic projection of the z-plane, on which $z = \infty$ is represented by a single point. If

$$z = a + t^{\mu}, \qquad u = Q(t), \qquad \text{or} \qquad z = t^{-\mu}, \qquad u = Q(t)$$

is the normal representation valid for $|t| < r$ of an irregular element of a given analytic form, then to each z_0 ($\neq a$ or ∞) which satisfies the condition

$$(3.1) \qquad |z - a| < r^{\mu} \qquad \text{or} \qquad |z| > r^{-\mu}$$

there correspond exactly μ regular function elements $u = \mathfrak{P}(z - z_0)$ with center z_0 which belong to the t-neighborhood $|t| < r$ of the irregular element. Briefly, we say that the irregular element "induces" μ regular elements at the point z_0. The inequality (3.1) determines a cap on the sphere, which I prefer to replace with the largest *cap* K inside it *whose center is* a (or ∞); let κ be the radius of K. As was shown above, two distinct irregular elements with the same center a or ∞ cannot induce the same regular function element at a point z_0 in their common convergence-cap. If $\mathfrak{e}_1$ and $\mathfrak{e}_2$ are two irregular elements with different centers and the associated caps are K_1 and K_2 with radii κ_1 and κ_2, then let k_1 and k_2 denote the caps with the same centers and radii $\frac{1}{2}\kappa_1$ and $\frac{1}{2}\kappa_2$; we assume that $\kappa_1 \geq \kappa_2$. If k_1 and k_2 intersect, let z_0 be a point of that intersection. I claim that no element induced by $\mathfrak{e}_1$ at z_0 is identical with any element induced by $\mathfrak{e}_2$ at z_0. For the center of K_2 is a point of K_1 and K_1 contains the great circle arc in K_2 which joins z_0 to the center of K_2. By continuation of a regular element at z_0, induced by $\mathfrak{e}_1$, along this arc to the center of K_2, we obtain one of the regular elements induced by $\mathfrak{e}_1$ at the center of K_2, and never their regular element $\mathfrak{e}_2$. Thus the claim is proved.

In the same way that we associated the (small) caps k_1 and k_2 with the irregular elements $\mathfrak{e}_1$ and $\mathfrak{e}_2$, we now associate a cap k with each irregular element $\mathfrak{e}$ of the given analytic form. If the irregular elements were not countable, then there would exist a rational point z_0 belonging to more than a countable number of caps k. Each irregular element associated with one of these caps induces at z_0 at least one regular element, with center z_0, of the analytic form; and, as we have just seen, different irregular elements induce different regular elements at z_0. But this is impossible, since the given analytic form contains no more than a countable number of regular elements with center z_0.

Thus the analytic form differs from the analytic function only in that a countable number of irregular elements have been added.

§ 4. The concept of a two-dimensional manifold

It was pointed out at the end of § 2 that one's intuitive grasp of an analytic form is greatly enhanced if one can represent each element of the form by a point on a surface $\mathfrak{F}$ in space in such a way that the representative points cover $\mathfrak{F}$ simply and so that every analytic chain of elements of the form becomes a continuous curve on $\mathfrak{F}$. To be sure, from a purely objective point of view, the problem of finding a surface to represent the analytic form in this visual way may be rejected as nonpertinent; for in essence, three-dimensional space has nothing to do with analytic forms, and one appeals to it not on logical-mathematical grounds, but because it is closely associated with our sense perception. To satisfy our desire for pictures and analogies in this fashion by forcing inessential representations on objects instead of taking them as they are could be called an anthropomorphism contrary to scientific principles. However, these reproaches of the pure logician are no longer pertinent if we pursue the other approach, already hinted at, in which the *analytic form is a two-dimensional manifold* to which all the ideas of continuity that we meet in ordinary geometry may be applied. To the contrary, not to use this approach is to overlook one of the most essential aspects of the topic.

Thus the concept "two-dimensional manifold" or "surface" will not be associated with points in three-dimensional space; rather it will be a much more general abstract idea. If any set of objects (which will play the role of points) is given and a continuous coherence between them, similar to that in the plane, is defined, then we shall speak of a two-dimensional manifold. Since all ideas of continuity may be reduced to the concept of neighborhood, two things are necessary to specify a two-dimensional manifold:

(1) to state what entities are the "points" of the manifold;

(2) to define the concept of "neighborhood."

More precisely, when will we say that a *two-dimensional manifold* $\mathfrak{F}$ *is given?* In the following circumstances:

(a) *There is given a set of objects called "points of the manifold* $\mathfrak{F}$." *For each point* $\mathfrak{p}$ *of the manifold* $\mathfrak{F}$, *certain subsets of* $\mathfrak{F}$ *are defined to be neighborhoods of* $\mathfrak{p}$ *on* $\mathfrak{F}$. *Each neighborhood of* $\mathfrak{p}$ *contains* $\mathfrak{p}$, *and for any pair of neighborhoods of* $\mathfrak{p}$ *there is a neighborhood of* $\mathfrak{p}$ *which is contained in each of the pair. If* $\mathfrak{U}_0$ *is a neighborhood of* $\mathfrak{p}_0$ *and if* $\mathfrak{p}$ *is in* $\mathfrak{U}_0$, *then there is a neighborhood of* $\mathfrak{p}$ *which is contained in* $\mathfrak{U}_0$. *If* $\mathfrak{p}_0$ *and* $\mathfrak{p}_1$ *are two distinct points of the manifold, then there exists a neighborhood of* $\mathfrak{p}_0$ *and a neighborhood of* $\mathfrak{p}_1$ *which are disjoint; that is, the two neighborhoods have no point in common.*

(b) *For each neighborhood* $\mathfrak{U}_0$ *of a given point* $\mathfrak{p}_0$ *of the manifold there is a one-to-one map of* $\mathfrak{U}_0$ *onto the interior,* K_0, *of a Euclidean circle* (say the unit disc $x^2 + y^2 < 1$ in the plane with Cartesian coordinates x and y) *with the following properties:*

(1) $\mathfrak{p}_0$ *corresponds to the center of the circle.*

(2) *If* $\mathfrak{p}$ *is any point of* $\mathfrak{U}_0$ *and if* $\mathfrak{U}$ *is a neighborhood of* $\mathfrak{p}$ *which consists only of points of* $\mathfrak{U}_0$, *then the image of* $\mathfrak{U}$ *in* K_0 *contains the image,* p, *of* $\mathfrak{p}$ *as an interior point; that is, there exists a disc* k *with center* p *such that each point of* k *is the image of a point of* $\mathfrak{U}$.

(3) *If* K *is a disc contained in* K_0, *with center* p, *then there exists a neighborhood* $\mathfrak{U}$ *of* $\mathfrak{p}$ *on* $\mathfrak{F}$ *whose image is contained in* K.

Part (a) defines the general concept of a manifold or topological space. Part (b) adds the restrictions characterizing two-dimensionality.[6]

The continuum of the real numbers (the straight line) is a manifold in our sense if for each positive ε we take the interval $|x - x_0| < \varepsilon$ as the ε-neighborhood of x_0. In the same way, each open interval $a < x < b$ on the line is a manifold: if x_0 belongs to this interval, then the inequality $|x - x_0| < \varepsilon$ defines a neighborhood of x_0, provided that $x_0 - \varepsilon > a$ and $x_0 + \varepsilon < b$. The Euclidean plane with Cartesian coordinates x, y, or the complex z-plane ($z = x + iy$) is a two-dimensional manifold if the interior of any circle with center p is a neighborhood of p. Also the interior K_0 of any circle in the z-plane is a two-dimensional manifold if one allows, as a neighborhood of p, the interior k of any circle with center p whose radius is small enough so that k is contained in K_0.

We now explain briefly how, with the concept of neighborhood, all the ideas of continuity in the ordinary plane can be carried over to arbitrary

---

[6]) For this reason part (a) includes statements that may be derived from the requirements in part (b).

two-dimensional manifolds. Given a point set $\mathfrak{E}$ on the surface, then the point $\mathfrak{p}$ belongs to the *hull*, $\mathfrak{E}'$, of $\mathfrak{E}$ if every neighborhood of $\mathfrak{p}$ contains points of $\mathfrak{E}$. Obviously $\mathfrak{E}$ is a part of $\mathfrak{E}'$. The set $\mathfrak{E}$ is called *closed* if $\mathfrak{E}'$ is the same as $\mathfrak{E}$. The hull of a set is always closed. A point $\mathfrak{p}$ of the set $\mathfrak{E}$ is an *isolated point* of $\mathfrak{E}$ if there exists a neighborhood of $\mathfrak{p}$ which contains no points of $\mathfrak{E}$ except $\mathfrak{p}$. The point $\mathfrak{p}$ is called an *interior point* of $\mathfrak{E}$ if there exists a neighborhood of $\mathfrak{p}$, all of whose points lie in $\mathfrak{E}$. A set $\mathfrak{E}$ is *open* if it consists exclusively of interior points. The *complement* $\bar{\mathfrak{E}}$ of a set $\mathfrak{E}$, which consists of all the points of the manifold which do not belong to $\mathfrak{E}$, is open if $\mathfrak{E}$ is closed, and vice versa. The interior points of a given set $\mathfrak{E}$ constitute its open *kernel;* the points of $\mathfrak{E}'$ which do not belong to the kernel of $\mathfrak{E}$ constitute the *boundary* of $\mathfrak{E}$.

If Σ is any collection, finite or infinite, of sets $\mathfrak{M}$ on $\mathfrak{F}$, then one can form the *union* and the *intersection.* A point belongs to the first if it is a point of at least one set $\mathfrak{M}$ of the system Σ; a point belongs to the intersection if it is a point of every set $\mathfrak{M}$ of the system. An intersection of closed sets is closed; the union of two, and hence of any finite number of, closed sets is also closed. A union of open sets is always open; the intersection of two, and hence of any finite number of, open sets is open.

If with each point $\mathfrak{p}$ of a certain set $\mathfrak{E}$ on $\mathfrak{F}$ there is associated a real or complex number $f(\mathfrak{p})$, then a *function* f is defined on $\mathfrak{E}$; the number $f(\mathfrak{p})$ is the *value* of this function at the point $\mathfrak{p}$. If $\mathfrak{p}_0$ is an arbitrary point of $\mathfrak{E}$, then f is said to be *continuous* at $\mathfrak{p}_0$ if for every positive number ε there exists a neighborhood $\mathfrak{U}_\varepsilon$ of $\mathfrak{p}_0$ on $\mathfrak{F}$ such that $|f(\mathfrak{p}) - f(\mathfrak{p}_0)| < \varepsilon$ for all $\mathfrak{p}$ that belong to both $\mathfrak{E}$ and $\mathfrak{U}_\varepsilon$.

A continuous *curve* γ on $\mathfrak{F}$ is given if with each real number λ of the unit interval $0 \leq \lambda \leq 1$ there is associated a point $\mathfrak{p}(\lambda)$ on $\mathfrak{F}$ such that the condition of contuinity is fulfilled. That is, if λ_0 is any value of the parameter λ and if $\mathfrak{U}_0$ is an arbitrary neighborhood of $\mathfrak{p}(\lambda_0)$ on $\mathfrak{F}$, then there exists a positive number ε such that $\mathfrak{p}(\lambda)$ is a point of $\mathfrak{U}_0$ for all λ in the unit interval which satisfy the condition $|\lambda - \lambda_0| \leq \varepsilon$. The curve γ *joins* the initial point $\mathfrak{p}(0)$ to the terminal point $\mathfrak{p}(1)$.

An open set $\mathfrak{E}$ is *connected,* and is called a *domain,* if every point in $\mathfrak{E}$ can be joined to every other point in $\mathfrak{E}$ by a curve in $\mathfrak{E}$. If $\mathfrak{p}_0$ is a point of $\mathfrak{E}$, then the set of points in $\mathfrak{E}$ which can be joined to $\mathfrak{p}_0$ by a curve in $\mathfrak{E}$ constitutes the *component* $\mathfrak{E}[\mathfrak{p}_0]$, which is determined by $\mathfrak{p}_0$. It is a domain, the maximal subdomain of $\mathfrak{E}$, which contains $\mathfrak{p}_0$. The components $\mathfrak{E}[\mathfrak{p}_0]$ and $\mathfrak{E}[\mathfrak{p}_1]$ determined by two points $\mathfrak{p}_0$ and $\mathfrak{p}_1$ of $\mathfrak{E}$ are either identical, namely in the case that $\mathfrak{p}_0$ and $\mathfrak{p}_1$ can be joined by a curve in $\mathfrak{E}$, or they have no point in common.

The concepts of continuous function and continuous curve are special cases of the general concept of a *continuous map*. Let $\mathfrak{F}_1$ and $\mathfrak{F}_2$ be any two manifolds, let $\mathfrak{E}_1$ be a point set on $\mathfrak{F}_1$, and $\mathfrak{E}_2$ on $\mathfrak{F}_2$. If with each point $\mathfrak{p}^{(1)}$ of $\mathfrak{E}_1$ there is associated a point $\mathfrak{p}^{(2)}$ of $\mathfrak{E}_2$, then a *map* S: $\mathfrak{p}^{(1)} \to \mathfrak{p}^{(2)}$ *of* $\mathfrak{E}_1$ *into* $\mathfrak{E}_2$ is given. The set of image points is called the *image* of $\mathfrak{E}_1$. The map S is said to be *continuous* if the following holds: if $\mathfrak{p}_0^{(1)}$ is any point of $\mathfrak{E}_1$, $\mathfrak{p}_0^{(2)}$ its image, and if $\mathfrak{U}_2$ is any neighborhood of $\mathfrak{p}_0^{(2)}$ on $\mathfrak{F}_2$, then there exists a neighborhood $\mathfrak{U}_1$ of $\mathfrak{p}_0^{(1)}$ on $\mathfrak{F}_1$ such that the image of any point which belongs to both $\mathfrak{E}_1$ and $\mathfrak{U}_1$ is a point of $\mathfrak{U}_2$. If $\mathfrak{E}_1$, $\mathfrak{E}_2$, $\mathfrak{E}_3$ are sets on the manifolds $\mathfrak{F}_1$, $\mathfrak{F}_2$, $\mathfrak{F}_3$ and if $S_1 : \mathfrak{p}_1 \to \mathfrak{p}_2 = \mathfrak{p}_1 S_1$ is a map of $\mathfrak{E}_1$ into $\mathfrak{E}_2$ and $S_2 : \mathfrak{p}_2 \to \mathfrak{p}_3 = \mathfrak{p}_2 S_2$ is a map of $\mathfrak{E}_2$ into $\mathfrak{E}_3$, then the successive application of these two maps, $\mathfrak{p}_1 \to \mathfrak{p}_3 = (\mathfrak{p}_1 S_1) S_2$, yields the map $S = S_1 S_2$, of $\mathfrak{E}_1$ into $\mathfrak{E}_3$. If S_1 and S_2 are continuous, then the composite map S is continuous. We discuss a particular case. If $\mathfrak{p}^{(1)}(\lambda)$ $(0 \leq \lambda \leq 1)$ is a continuous curve on $\mathfrak{F}_1$, contained in a subset $\mathfrak{E}_1$ which is mapped continuously into a second manifold $\mathfrak{F}_2$, then one can define a continuous curve on $\mathfrak{F}_2$ as follows. To each value λ of the parameter there corresponds the point $\mathfrak{p}^{(2)}$ on $\mathfrak{F}_2$ which is the image of $\mathfrak{p}^{(1)}(\lambda)$ under the map of $\mathfrak{E}_1$ into $\mathfrak{F}_2$. This curve is the image of the initial curve.

When discussing a *one-to-one map* S of a set $\mathfrak{E}_1$ onto a set $\mathfrak{E}_2$, it is useful to consider the pair consisting of S and its inverse S'. One has then a map S of $\mathfrak{E}_1$ onto $\mathfrak{E}_2$ and a map S' of $\mathfrak{E}_2$ onto $\mathfrak{E}_1$, as well as the map SS' which is the identity map carrying each point $\mathfrak{p}_1$ of $\mathfrak{E}_1$ into itself, and also the identity map $S'S$ of $\mathfrak{E}_2$ onto itself. The inverse S' is customarily denoted by S^{-1}. If both S and S' are continuous, we talk of a *bicontinuous map* S. If S_1 is a one-to-one map of $\mathfrak{E}_1$ onto $\mathfrak{E}_2$ and if S_2 is a one-to-one map of $\mathfrak{E}_2$ onto $\mathfrak{E}_3$, then the composite map $S = S_1 S_2$ of $\mathfrak{E}_1$ onto $\mathfrak{E}_3$ is one-to-one and its inverse is $S^{-1} = S_2^{-1} S_1^{-1}$. (Order of the factors reversed! Since in dressing the shirt comes first and the coat last, the order is reversed in undressing.) The product of two bicontinuous maps S_1 and S_2 is also bicontinuous.

Along with the concept of a continuous map, which is used primarily for closed sets, there appears a concept of continuity which is suited to open sets and for which I propose the label "neighborhood-true" (*umgebungstreu*).[7] The map S of an open set $\mathfrak{E}_1$ on $\mathfrak{F}_1$ into an open set $\mathfrak{E}_2$ on $\mathfrak{F}_2$ is *neighborhood-true* if for every neighborhood $\mathfrak{U}^{(1)}$ of any point $\mathfrak{p}_0^{(1)}$ of $\mathfrak{E}_1$, the image, $\mathfrak{p}_0^{(2)}$, of $\mathfrak{p}_0^{(1)}$ is an interior point of the image, $\mathfrak{V}^{(2)}$, of $\mathfrak{U}^{(1)}$. If S is a one-to-one map of $\mathfrak{E}_1$ onto $\mathfrak{E}_2$, then we consider S^{-1} also. If S is neighborhood-true, then S^{-1} is continuous, and vice versa. If S and S^{-1} are both neighborhood-

---

[7]) "Domain-continuous" ["gebiets-stetig"] in the first edition.

true, then the map is called *reversibly neighborhood-true.* Thus for open sets, the concepts reversibly neighborhood-true and bicontinuous agree. Such maps are commonly called *topological.* The conditions 1, 2, 3, above, on the neighborhoods in the definition of a two-dimensional manifold may now be formulated simply as follows: any neighborhood of $\mathfrak{p}_0$ possesses a topological map on a plane disc K_0 such that $\mathfrak{p}_0$ corresponds to the center of K_0.

Today[8] a set $\mathfrak{E}$ on $\mathfrak{F}$ is called *compact* if it has the following property. Let a neighborhood $\mathfrak{U}(\mathfrak{p})$ be associated with each point $\mathfrak{p}$ in $\mathfrak{E}$; then one can choose a finite number of points $\mathfrak{p}_1, \dots, \mathfrak{p}_N$, of $\mathfrak{E}$ such that the associated neighborhoods $\mathfrak{U}(\mathfrak{p}_1), \dots, \mathfrak{U}(\mathfrak{p}_N)$ cover $\mathfrak{E}$ completely "like roof tiles" (as Felix Klein used to say); that is, so that every point of $\mathfrak{E}$ lies in one of these neighborhoods. *The continuous image of a compact set is compact. Every closed subset $\mathfrak{C}$ of a compact set $\mathfrak{E}$ is compact.* For let there be given, for each $\mathfrak{p}$ in $\mathfrak{C}$, a neighborhood $\mathfrak{U}(\mathfrak{p})$ on $\mathfrak{F}$. Since the complement $\bar{\mathfrak{C}}$ is an open set, one can associate with each point $\mathfrak{p}$ of $\mathfrak{E}$ which does not belong to $\mathfrak{C}$ a neighborhood $\mathfrak{U}(\mathfrak{p})$ on $\mathfrak{F}$ which is contained in $\bar{\mathfrak{C}}$. Since $\mathfrak{E}$ is compact, there are finitely many points $\mathfrak{p}_1, \dots, \mathfrak{p}_N$ of $\mathfrak{E}$ such that the associated neighborhoods $\mathfrak{U}(\mathfrak{p}_i)$ $(i = 1, \dots, N)$ cover $\mathfrak{E}$. If we omit the $\mathfrak{p}_i$ which do not belong to $\mathfrak{C}$, then the neighborhoods associated with the remaining $\mathfrak{p}_i$ cover $\mathfrak{C}$; for if $\mathfrak{p}_i$ is not a point of $\mathfrak{C}$, then $\mathfrak{U}(\mathfrak{p}_i)$ is disjoint from $\mathfrak{C}$. A consequence of this theorem is the fact that *the intersection of any system of compact sets is compact.* For this intersection is a closed subset of a compact set. *A compact set $\mathfrak{E}$ which consists entirely of isolated points contains only a finite number of points.* By associating with each point $\mathfrak{p}$ of this set a neighborhood $\mathfrak{U}(\mathfrak{p})$ which contains no other point of the set, and by choosing finitely many points $\mathfrak{p}_1, \dots, \mathfrak{p}_N$, whose associated neighborhoods $\mathfrak{U}(\mathfrak{p}_i)$ cover $\mathfrak{E}$, we see that the complete set is exhausted by these points.

It is a not wholly trivial *property of compact sets that they are closed.* Let $\mathfrak{E}$ be a compact set, and $\mathfrak{a}$ a point not belonging to $\mathfrak{E}$. For the point $\mathfrak{a}$ of the complementary set $\bar{\mathfrak{E}}$ we must construct a neighborhood of $\mathfrak{a}$ which is contained in $\bar{\mathfrak{E}}$. If this is done, no matter what point of $\bar{\mathfrak{E}}$ $\mathfrak{a}$ is, then it is proved that $\bar{\mathfrak{E}}$ is open and hence $\mathfrak{E}$ is closed. This is the construction: for each point $\mathfrak{p}$ of $\mathfrak{E}$ one can find neighborhoods, $\mathfrak{U}(\mathfrak{p})$ of $\mathfrak{p}$ and $\mathfrak{V}_\mathfrak{p}$ of $\mathfrak{a}$, which are disjoint; for $\mathfrak{p} \neq \mathfrak{a}$. Because of the compactness there are finitely many of the points $\mathfrak{p}, \mathfrak{p}_1, \dots, \mathfrak{p}_N$, whose associated neighborhoods $\mathfrak{U}(\mathfrak{p}_i)$

---

[8]) For some time the term bicompact was used, since the term compact was used for a wider class of sets. Today the terminology accepted here has been generally agreed upon, at least in America.

($i = 1, \dots, N$) cover $\mathfrak{E}$. There exists a neighborhood $\mathfrak{B}(\mathfrak{a})$ of $\mathfrak{a}$ which is contained in each of the finitely many neighborhoods $\mathfrak{B}_{\mathfrak{p}_i}$ of $\mathfrak{a}$ ($i = 1, \dots, N$). It is disjoint from all the N-neighborhoods $\mathfrak{U}(\mathfrak{p}_i)$, hence from the set $\mathfrak{E}$ which they cover, and thus indeed lies completely in $\bar{\mathfrak{E}}$.

The theorem of Weierstrass, that *a function* $f(\mathfrak{p})$, *defined and continuous on a compact set* $\mathfrak{E}$, *assumes a maximum and a minimum,* is a special case of the fact that the continuous image of a compact set is compact. I repeat the proof. With each point $\mathfrak{p}$ of $\mathfrak{E}$ we can associate a neighborhood $\mathfrak{U}(\mathfrak{p})$ such that the values of f in this neighborhood differ from the value $f(\mathfrak{p})$ at $\mathfrak{p}$ itself by less than ± 1. By covering $\mathfrak{E}$ with finitely many of these neighborhoods, it follows that f is bounded; from the fact that the image of $\mathfrak{E}$ by f on the real axis is closed, it follows that the least upper and greatest lower bounds of the set of values of f belong themselves to the set of values.

The content of the Heine–Borel theorem[9] is: a closed interval $a \leq x \leq a'$ in the continuum of real numbers x (a and a' any two numbers, $a \leq a'$) is compact. The argument used to prove this theorem also yields the theorem: every rectangle in the x, y-plane ("two-dimensional interval") defined by the inequalities

$$a \leq x \leq a', \qquad b \leq y \leq b'$$

(a, a', b, b' any real numbers, $a' \geq a$, $b' \geq b$) is compact. Consequently it follows from one of the theorems above that *every closed bounded set in the plane is compact.* For a set in the plane is called *bounded* if it is contained in some rectangle. Conversely, a compact subset of the plane is necessarily bounded. For with each point p of the set associate the neighborhood consisting of the interior of the circle with center p and radius 1; a set which is covered by a finite number of such discs is bounded.

From the fact that the closed unit interval $0 \leq \lambda \leq 1$ is compact we get the following immediate consequence: if to each value λ_0 of this interval there corresponds an open interval $\mathfrak{u}(\lambda_0)$ containing λ_0, then one can split the unit interval into finitely many subintervals by points $\lambda_0 = 0 < \lambda_1 < \lambda_2 < \cdots < \lambda_{n-1} < \lambda_n = 1$ so that each subinterval is contained in one neighborhood $\mathfrak{u}(\lambda)$ associated with one point λ of the subinterval. This carries over for a continuous curve γ, $\mathfrak{p} = \mathfrak{p}(\lambda)$, on a manifold $\mathfrak{F}$ as follows. Let there be associated with each point $\mathfrak{p}$ of $\mathfrak{F}$ a neighborhood $\mathfrak{U}(\mathfrak{p})$, or perhaps only with each point λ of the curve a neighborhood $\mathfrak{U}[\lambda]$ of $\mathfrak{p} = \mathfrak{p}(\lambda)$. Then this curve may be split into a finite number of subarcs, each containing one distinguished point λ, such that each subarc is contained in the neighborhood associated with the distinguished point of the subarc

[9]) See, for example, Lebesgue, *Leçons sur l'intégration,* Paris 1904, 104–105.

("*standard subdivision* of γ"). [Instead of a neighborhood $\mathfrak{U}(\mathfrak{p})$ any set such that $\mathfrak{p}$ is an interior point will do.]

We use this to replace the definition of domain by a conceptually simpler one. Namely, we claim: *an open set $\mathfrak{E}$ is connected if and only if it cannot be split into two open subsets $\mathfrak{E}_1$, $\mathfrak{E}_2$ except when one of them is void.* Here *split* is to be understood in the strong sense, that every point of $\mathfrak{E}$ belongs to one and to only one of the two subsets $\mathfrak{E}_1$, $\mathfrak{E}_2$. In fact, if the set is not connected in our earlier sense, we choose a point $\mathfrak{p}_0$ in $\mathfrak{E}$, and separate those points of $\mathfrak{E}$ which can be joined to $\mathfrak{p}_0$ by a continuous curve in $\mathfrak{E}$ from the other points of $\mathfrak{E}$. That is a splitting of $\mathfrak{E}$ into two nonvoid open sets $\mathfrak{E}_1$, $\mathfrak{E}_2$. Conversely, given such a splitting, it is impossible to join a point $\mathfrak{a}_1$ of $\mathfrak{E}_1$ to a point $\mathfrak{a}_2$ of $\mathfrak{E}_2$ by a continuous curve. For, associate with each point $\mathfrak{p}$ of $\mathfrak{E}_1$ a neighborhood contained in $\mathfrak{E}_1$ and with each point $\mathfrak{p}$ of $\mathfrak{E}_2$ a neighborhood contained in $\mathfrak{E}_2$. Then one gets a subdivision of the curve, by interpolated division points, into finitely many arcs, each of which lies in one of the specified neighborhoods. By proceeding from division point to division point, one finds that all lie in $\mathfrak{E}_1$ since the first, $\mathfrak{a}_1$, lies in $\mathfrak{E}_1$; a contradiction of the assumption that the last, $\mathfrak{a}_2$, is located in $\mathfrak{E}_2$ rather than $\mathfrak{E}_1$.

A set $\mathfrak{E}$ on $\mathfrak{F}$ will be called *scattered*[10] (in or on $\mathfrak{F}$), if for each point $\mathfrak{p}$ of $\mathfrak{F}$ a neighborhood $\mathfrak{U}(\mathfrak{p})$ can be determined so that no points of $\mathfrak{U}(\mathfrak{p})$, distinct from $\mathfrak{p}$, belong to $\mathfrak{E}$. For example, the poles of a function, regular analytic except for poles in a domain $\mathfrak{G}$ of the complex z-plane, is a scattered set on $\mathfrak{G}$. A scattered set on $\mathfrak{F}$ can also be described as a closed set consisting of isolated points.

Let $\mathfrak{F}$ and $\mathfrak{F}'$ be manifolds, and let $\mathfrak{E}$ be a compact set on $\mathfrak{F}$. Given a continuous map, $\mathfrak{p} \to \mathfrak{p}' = \phi(\mathfrak{p})$, of $\mathfrak{E}$ into $\mathfrak{F}'$, then, as we saw, the image $\mathfrak{E}'$ is also compact. If distinct points $\mathfrak{p}$ of $\mathfrak{E}$ always corresponds to distinct points $\mathfrak{p}'$ of $\mathfrak{E}'$, then one can consider the inverse map, $\mathfrak{p}' \to \mathfrak{p} = \psi(\mathfrak{p}')$. I claim that *this map is also continuous,* so that we have a bicontinuous map of $\mathfrak{E}$ onto $\mathfrak{E}'$. For, let $\mathfrak{a}$ be a point of $\mathfrak{E}$ and let $\mathfrak{U}_0$ be a neighborhood of $\mathfrak{a}$ (on $\mathfrak{F}$). Those points of $\mathfrak{E}$ which do not lie in $\mathfrak{U}_0$ form a closed subset $\mathfrak{E}_1$ of $\mathfrak{E}$. Therefore $\mathfrak{E}_1$, as $\mathfrak{E}$ itself, is compact; also the continuous ϕ-image, $\mathfrak{E}_1'$, is compact. Consequently $\mathfrak{E}_1'$ is closed and, since it does not contain the point $\mathfrak{a}' = \phi(\mathfrak{a})$, there exists a neighborhood $\mathfrak{V}_0$ of $\mathfrak{a}'$ which has no point in common with $\mathfrak{E}_1'$. The ψ-image $\mathfrak{p} = \psi(\mathfrak{p}')$ of a point $\mathfrak{p}'$, of $\mathfrak{E}'$, which lies in $\mathfrak{V}_0$ is not in $\mathfrak{E}_1$ and hence lies in $\mathfrak{U}_0$. With that we have proved: If $\mathfrak{a}'$ is

---

[10]) With this word [verstreut] I have called to mind a beautiful poem, *Sommerabend,* of Hermann Hesse, containing the lines:

> Sommernacht hat ihre dünnen Sterne verstreut,
> Jugendgedächtnis duftet im mondhellen Laub...

an arbitrary point of $\mathfrak{E}'$, $\mathfrak{a} = \psi(\mathfrak{a}')$, and if $\mathfrak{U}_0$ is any neighborhood of $\mathfrak{a}$, then there exists a neighborhood $\mathfrak{V}_0$ of $\mathfrak{a}'$ whose ψ-image lies in $\mathfrak{U}_0$. But this is exactly what the continuity of ψ means.

The definition of what is to be regarded as a "neighborhood" of a point on a manifold $\mathfrak{F}$ can be altered within limits without our regarding the manifold as changed. Suppose that the objects which serve as points of $\mathfrak{F}$ are unchanged, but that we replace the original ("first") specification of neighborhood by a second which also satisfies the conditions above. If the following result is valid: for any point $\mathfrak{p}_0$ on $\mathfrak{F}$, any second-definition neighborhood of $\mathfrak{p}_0$ contains a first-definition neighborhood of $\mathfrak{p}_0$ and conversely, then we agree that the manifold determined by the second definition is to be regarded as no different from that determined by the first. In this agreement there appears again a "definition by abstraction," the verification of which we can leave to the reader. Also, none of the ideas of continuity listed above is affected in any way by replacing first-definition neighborhoods by second-definition neighborhoods: a set which is closed or compact or a domain according to the first definition has the same property according to the second definition; similarly with continuous functions, continuous curves, continuous maps, etc.

That branch of mathematics which deals with the continuity properties of two- (and more) dimensional manifolds is called *analysis situs* or *topology*. This discipline has played an important role in function theory since the time of Riemann; in the following sections we must concern ourselves more deeply with the topology of two-dimensional manifolds. Two manifolds must be regarded as *equivalent in the topological sense* if they can be mapped point for point in a reversibly neighborhood-true (topological) fashion on each other. One may regard each manifold in an equivalence class of two-dimensional manifolds as a realization of one and the same "ideal manifold" in which the individual features which distinguish the different realizations are expunged (definition by abstraction). The ideas of continuity collected above are pure analysis-situs concepts, for they remain invariant under topological maps. By regarding the analytic form as a two-dimensional manifold we are led to placing the analysis-situs properties of this manifold in the foreground as the most incisive and primitive properties. This is the way Riemann proceeded in his construction of function theory.

We now narrow the concept of a two-dimensional manifold by imposing two further conditions. The first is that the manifold $\mathfrak{F}$ itself is to be a *domain,* in other words, it is to consist of one piece. The second postulate (axiom of countability) is deeper: *if with each point* $\mathfrak{p}$ *there is associated a neighborhood* $\mathfrak{U}(\mathfrak{p})$, *then a countable sequence of points* $\mathfrak{p}_1, \mathfrak{p}_2, \ldots$ *can be chosen so that the associated neighborhoods* $\mathfrak{U}(\mathfrak{p}_1)$, $\mathfrak{U}(\mathfrak{p}_2)$, ... *cover* $\mathfrak{F}$ *completely*. For two-dimensional manifolds which satisfy both conditions, we use mostly the shorter term *surface*. The surface is *compact* if, in the second condition, the term "countable sequence" can be replaced by "finite set." In the older literature it was customary to call compact surfaces *closed surfaces*. I shall remain loyal to this usage, but, to avoid confusion with the concept of open set introduced above, I shall not use the word *open* in the sense of unclosed = noncompact.

A subdomain $\mathfrak{G}$ of a two-dimensional manifold $\mathfrak{F}$ is itself a two-dimensional manifold, $\boldsymbol{\mathfrak{G}}$, if we allow as a neighborhood on $\boldsymbol{\mathfrak{G}}$ of a point of $\mathfrak{G}$ any neighborhood of the point on $\mathfrak{F}$ which is contained in $\mathfrak{G}$.

Then any continuous curve on $\mathfrak{F}$ which lies completely in $\mathfrak{G}$ is a continuous curve on $\boldsymbol{\mathfrak{G}}$, etc. Thus, from now on we can dispense with the pedantic distinction between $\mathfrak{G}$ and $\boldsymbol{\mathfrak{G}}$, without danger of confusion. Any subset of $\mathfrak{G}$ is a subset of $\mathfrak{F}$. If it is open in $\mathfrak{G}$, it is open in $\mathfrak{F}$, and conversely. To this extent, the concept of open set is *absolute*; that is, it is independent of whether we regard a set as a subset of $\mathfrak{F}$ or of $\mathfrak{G}$. This is not the case with the concept of closed set. For example, the interior, $\mathfrak{E}$, of the unit circle is not closed as a subset of the plane, but it is closed as a subset of itself. For this reason it is noteworthy that the concept of compactness is obviously absolute in the sense laid down.

It is important to observe that *a subdomain* $\mathfrak{G}$ *of a surface* $\mathfrak{F}$ *is also a surface*. To see this it is enough to show that $\mathfrak{G}$ may be covered by a countable collection G_1, G_2, ... of compact subsets of $\mathfrak{G}$. For if a neighborhood $\mathfrak{U}(\mathfrak{p})$, contained in $\mathfrak{G}$, is given for each point $\mathfrak{p}$ of $\mathfrak{G}$, then in each G_i there are finitely many points, $\mathfrak{p}_i'$, $\mathfrak{p}_i''$, ... , $\mathfrak{p}_i^{(n_i)}$ whose associated neighborhoods cover G_i. Then the neighborhoods associated with the points

$$\mathfrak{p}_i', \mathfrak{p}_i'', \ldots, \mathfrak{p}_i^{(n_i)} \qquad (i = 1, 2, \ldots)$$

cover $\mathfrak{G}$.

We now proceed as follows. Let $\mathfrak{U}(\mathfrak{a})$ be a neighborhood of the point $\mathfrak{a}$ on $\mathfrak{F}$ and let A be a topological map of $\mathfrak{U}(\mathfrak{a})$ onto the interior E of the unit circle in the xy-plane. We replace E by a closed square Q, contained in E, which is defined by

$$(4.1) \qquad -\tfrac{1}{2} \leq x \leq \tfrac{1}{2}, \qquad -\tfrac{1}{2} \leq y \leq \tfrac{1}{2}.$$

(When we are concerned with covering the plane with elementary pieces, squares are preferable to circles, for they are easier to cut up.) It is also useful to introduce the number $\max(|x_2 - x_1|, |y_2 - y_1|)$ as the distance between the two points (x_1, y_1), (x_2, y_2). The *elementary subdivision* of a square Q is the subdivision into four squares of equal size by parallels to the sides (coordinate axes).

By the map A, Q, specified by (4.1), is the image of a set Q on $\mathfrak{F}$; this set is, like the square itself, compact and it contains a neighborhood $\mathfrak{u}(\mathfrak{a})$ of $\mathfrak{a}$. Having constructed a "square" Q with center $\mathfrak{a}$ and a neighborhood $\mathfrak{u}(\mathfrak{a})$ contained in Q for each point $\mathfrak{a}$ of $\mathfrak{F}$, one can find countably many points $\mathfrak{p}_i$ ($i = 1, 2, \ldots$), such that $\mathfrak{F}$ is covered by the associated neighborhoods $\mathfrak{u}(\mathfrak{p}_i)$ and, a fortiori, by the associated squares $Q = Q_i$. If $\mathfrak{G}$ is the given domain on $\mathfrak{F}$, retain those squares Q_i which are contained in $\mathfrak{G}$; they form a sequence $Q_1', Q_2', \ldots$. Apply an elementary subdivision to each of the other squares Q_i and retain those of the new half-size squares which are contained in $\mathfrak{G}$: $Q_1'', Q_2'', \ldots$. The leftover squares are again subjected to an elementary subdivision, and the new squares which are contained in $\mathfrak{G}$ are sifted out, and so on. Certainly we obtain in this way a complete covering of $\mathfrak{G}$ by compact "elementary squares" $Q_i^{(\nu)}$ in $\mathfrak{G}$. For let $\mathfrak{p}$ be any point of $\mathfrak{G}$; $\mathfrak{p}$ belongs to one of the "large" squares $Q_i = Q$. This Q is the part of $\mathfrak{U}(\mathfrak{p}_i)$ which corresponds to (4.1) under a topological map A of $\mathfrak{U}(\mathfrak{p}_i)$ onto the disc E. Since $\mathfrak{p}$ is an interior point of $\mathfrak{G}$, there exists a natural number $n \geq 2$ large enough so that all points whose distance from $\mathfrak{p}$ [in the image of $\mathfrak{U}(\mathfrak{p}_i)$ by A] is $\leq 2^{-n}$ will lie in $\mathfrak{G}$. Apply in succession n elementary subdivisions to Q; the resulting squares have sides of length 2^{-n}. Then one (or possibly those two or four) of the small squares which contains $\mathfrak{p}$ will be contained in $\mathfrak{G}$. Thus the point $\mathfrak{p}$ will be picked up in one of the elementary squares $Q_i^{(\nu)}$ no later than the nth step; that is, for $\nu \leq n + 1$. As a consequence of the numbering of the elementary squares by the double index (ν, i), these may be arranged in a countable sequence.

§ 5. Examples of surfaces

EXAMPLE 1. The *Euclidean plane* (not closed).

EXAMPLE 2. The *interior of a square* (not closed).

Figures 1 and 2 indicate particular *triangulations* of these surfaces. That is, a particularly regular covering of the surface with countably many compact pieces (triangles) which have no interior points in common. Such triangulations play a decisive role in combinatorial analysis situs and were used to the full in the first edition of this book. But here we can dispense with them entirely.

EXAMPLE 3. The *spherical surface* in three-dimensional Euclidean space with Cartesian coordinates ξ_1, ξ_2, ξ_3 consists of the points (ξ_1, ξ_2, ξ_3), which satisfy the equation $\xi_1^2 + \xi_2^2 + \xi_3^2 = 1$. The neighborhood of radius ρ of the point $P^0 = (\xi_1^0, \xi_2^0, \xi_3^0)$ shall consist of all points P whose angular distance from P^0 is $< \rho$; here ρ is any positive number $< \pi/2$. By orthogonal projection onto the tangent plane at P^0, this neighborhood is mapped

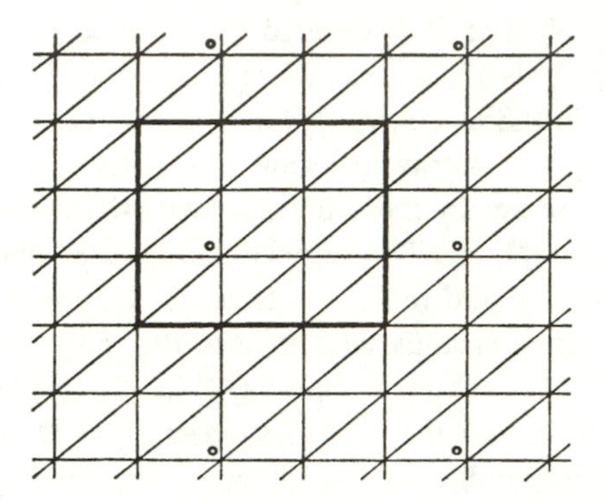

Fig. 1. Triangulation of the plane and the torus.

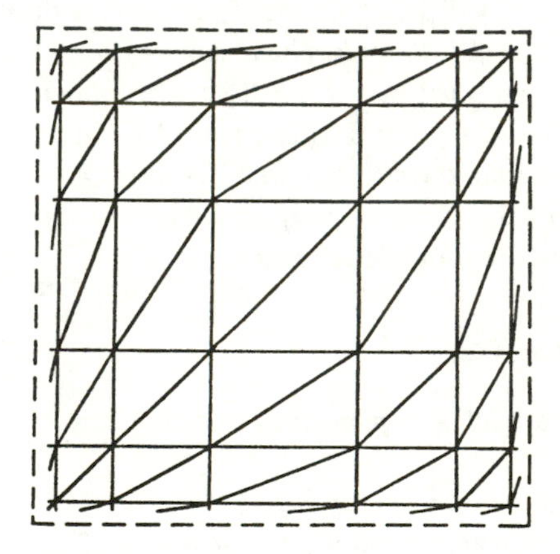

Fig. 2. Triangulation of the interior of a square.

topologically onto a plane disc with center P^0 and radius $r = \sin \rho$. The neighborhood may also be characterized by the inequality

$$(5.1) \qquad (\xi_2\xi_3^0 - \xi_3\xi_2^0)^2 + (\xi_3\xi_1^0 - \xi_1\xi_3^0)^2 + (\xi_1\xi_2^0 - \xi_2\xi_1^0)^2 < r^2 .$$

The sphere is a closed surface. For, split it into northern and southern hemispheres and project these orthogonally onto the plane of the equator; thus the two hemispheres appear as one-to-one continuous images of the closed unit disc, which is compact. Therefore each hemisphere is itself compact and thus the sphere is also.

EXAMPLE 4. The *projective plane* can be defined arithmetically as the set of all ratios $\xi_1 : \xi_2 : \xi_3$ obtained from three real numbers $(\xi_1, \xi_2, \xi_3) \neq (0,0,0)$. It can also be defined as the continuum of lines through the origin in three-dimensional Euclidean space with the Cartesian coordinates ξ_i. Every such line meets the sphere in two antipodal points, (ξ_1, ξ_2, ξ_3) and $(-\xi_1, -\xi_2, -\xi_3)$. Therefore one can say that the projective plane arises from the sphere by identification of each pair of antipodal points. Consonant with (5.1), the

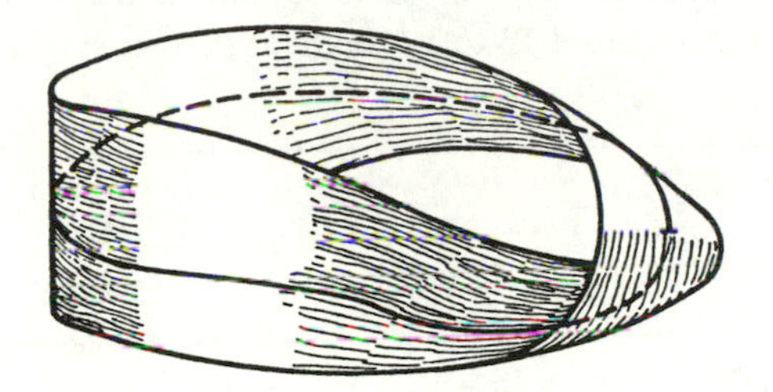

Fig. 3. The Möbius strip.

neighborhood of radius r $(0 < r < 1)$ of the point $\xi_1^0 : \xi_2^0 : \xi_3^0$ will consist of all points $\xi_1 : \xi_2 : \xi_3$ of the projective plane for which

$$\frac{(\xi_2\xi_3^0 - \xi_3\xi_2^0)^2 + (\xi_3\xi_1^0 - \xi_1\xi_3^0)^2 + (\xi_1\xi_2^0 - \xi_2\xi_1^0)^2}{[\xi_1^2 + \xi_2^2 + \xi_3^2][(\xi_1^0)^2 + (\xi_2^0)^2 + (\xi_3^0)^2]} < r^2 .$$

EXAMPLE 5. The *Möbius strip*[11] (Fig. 3) arises when one takes a long narrow rectangle, twists it through 180° in three-dimensional space, and bends it around to identify the short sides in opposite directions. This surface in space, referred to rectangular coordinates xyz, is most simply represented in terms of two real parameters ρ, ϕ as follows:[12]

$$x = (a - \rho \sin \tfrac{1}{2}\phi) \cos \phi,$$
$$y = (a - \rho \sin \tfrac{1}{2}\phi) \sin \phi,$$
$$z = \rho \cos \tfrac{1}{2}\phi,$$

where ϕ is unrestricted and ρ is subject to the condition $|\rho| < h$; a and h are

---

11) Möbius, *Werke*, **2**, 484–485 and 519–521.

12) The surface represented by these equations is by no means a developable.

positive constants, of which h must be the smaller. In the sense of analysis situs, the Möbius strip is obviously identical with the following manifold $\mathfrak{B}$.

In a Euclidean plane with rectangular coordinates ρ and ϕ consider the strip $|\rho| < 1$ and the discrete group Γ of paddle motions[13] $(\rho, \phi) \rightarrow (\rho', \phi')$ defined by

$$(5.2) \qquad \rho' = (-1)^n \rho, \qquad \phi' = \phi + 2n\pi \qquad (n = 0, \pm 1, \pm 2, \ldots).$$

A point set S in this plane is called a *system of equivalent points with respect to* Γ (more briefly: a *point system of* Γ) if points of S go into points of S under each motion in Γ and any point of S can be carried into any other by some motion in Γ. A point system of Γ which is contained in the strip $|\rho| < 1$ will be a "point" of the manifold $\mathfrak{B}$ to be defined. If S_0 is such a "point" of $\mathfrak{B}$

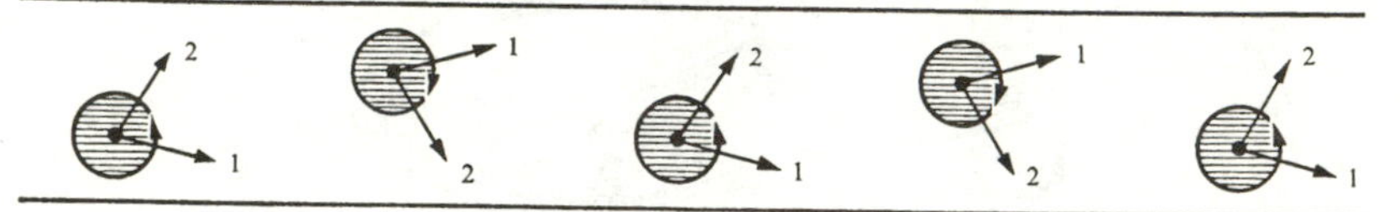

Fig. 4. Point of $\mathfrak{B}$ with neighborhood and angle.

and if one places congruent circles about each (Euclidean) point of the system S_0, which circles are contained in the strip and do not overlap, then those point systems S of Γ which are represented by a single point in each of these circles form a "neighborhood" of S_0 on $\mathfrak{B}$. If one says that an ordinary point which belongs to the point system S of Γ *lies over* the "point" S of $\mathfrak{B}$, then the parallel strip appears as a *covering surface* over $\mathfrak{B}$ which covers $\mathfrak{B}$ with infinitely many sheets, without, however, being *branched* relative to $\mathfrak{B}$ at any point. In the parallel strip we use the Euclidean measure of angles; by regarding the strip as a covering of $\mathfrak{B}$, this angular measure is carried over to $\mathfrak{B}$ immediately. In the process, however, the angle becomes ambiguous, not merely through integral multiples of 2π, but it also loses its sign (Fig. 4). This circumstance is related to the fact that the Möbius strip has the property that one calls "one-sidedness"; that is, without leaving the surface or climbing over an edge, one can get from one side of it to the other. Because of its function theoretic significance we shall examine this property more precisely later.

[13]) The term "paddle motion" ("Paddelbewegung", a Low German colloquialism) is meant to indicate that the group is obtained as follows: one repeatedly flips the plane over about an axis, alternately to the left and to the right, and at each flip pushes the plane forward (through a distance 2π) in the direction of that axis.

EXAMPLE 6. The *torus* (Fig. 5) arises when one rotates a circle of radius r about an axis which lies in the plane of the circle and which does not intersect the circle. Let $R\,(> r)$ be the perpendicular distance of the center of the circle from the axis. The rectangular coordinates xyz of points on the surface may be expressed with the aid of two real parameters σ, ϕ, whose geometrical significance is obvious, as follows:

$$x = (R + r\cos\sigma)\cos\phi,$$
$$y = (R + r\cos\sigma)\sin\phi,$$
$$z = r\sin\sigma.$$

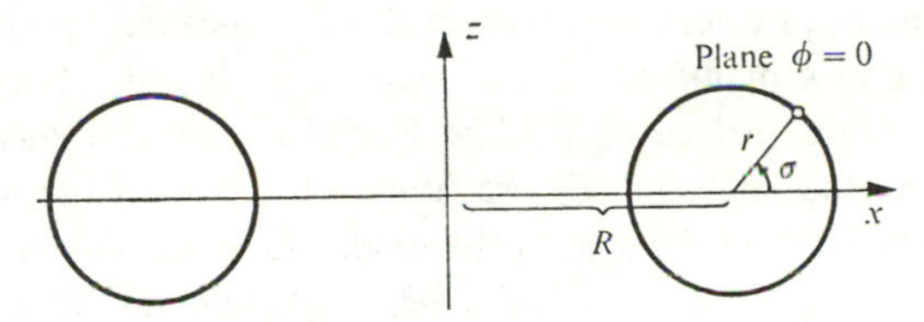

Fig. 5. Construction of the torus.

As a surface in three-dimensional Euclidean space, there is a natural (Euclidean) angle measure on the torus. If a curve

$$\phi = \phi(\lambda), \qquad \sigma = \sigma(\lambda)$$

is given on the torus, such that $d\phi/d\lambda$ and $d\sigma/d\lambda$ are continuous and $(d\phi/d\lambda)^2 + (d\sigma/d\lambda)^2 \neq 0$, then the differential quotient of the arc length $s = s(\lambda)$ along the curve may be computed from

$$\left(\frac{ds}{d\lambda}\right)^2 = (R + r\cos\sigma)^2\left(\frac{d\phi}{d\lambda}\right)^2 + r^2\left(\frac{d\sigma}{d\lambda}\right)^2.$$

If one replaces σ with the new parameter ψ,

$$\psi = \int_0^\sigma \frac{d\sigma}{R/r + \cos\sigma}, \tag{5.3}$$

then one can write

$$ds^2 = e(d\phi^2 + d\psi^2), \qquad \text{where } e = (R + r\cos\sigma)^2. \tag{5.4}$$

By (5.3), ψ is a single-valued continuously differentiable function of σ; σ, as a function of ψ, has these same properties. The values (ϕ, ψ), which correspond to one point of the torus, are determined only to within integral multiples of

$$a = 2\pi, \qquad b = \int_0^{2\pi} \frac{d\sigma}{R/r + \cos\sigma} = \frac{2\pi r}{\sqrt{R^2 - r^2}} .$$

Thus we are led to consider the group of translations

$$\left.\begin{array}{ll} \phi' = \phi + ma, & \psi' = \psi + nb \\ m = 0, \pm 1, \pm 2, \ldots & \\ n = 0, \pm 1, \pm 2, \ldots & \end{array}\right\} \Gamma$$

in the Euclidean plane with rectangular coordinates ϕ, ψ, and to regard a system of points equivalent with respect to Γ, a so-called "point lattice," as a "point" of a new manifold $\mathfrak{T}$. The plane is an infinite sheeted, nowhere branched, covering surface of $\mathfrak{T}$. The Euclidean angular measure carries over directly to the manifold $\mathfrak{T}$, and (unlike the case of the Möbius strip) there is no ambiguity in the sign of the angle. As in the $\phi\psi$-plane itself, the neighborhood of radius ε of the point (ϕ_0, ψ_0) is defined by the inequality $(\phi - \phi_0)^2 + (\psi - \psi_0)^2 < \varepsilon^2$; only now the positive radius ε is restricted by the condition $\varepsilon < \min(a, b)$. By its representation in terms of the parameters ϕ, ψ, the torus is mapped topologically onto $\mathfrak{T}$; but in addition, the map is also *conformal*. For the equation (5.4) says that the ratio of the element of arc, ds, on the torus to the image element $(\sqrt{d\phi^2 + d\psi^2})$ in the $\phi\psi$-plane is $\sqrt{e} = R + r\cos\sigma$; thus the ratio depends only on the point at which the element of arc is located, not on its direction. This conformal map is the foundation for the theory of analytic functions on the torus. The torus is *closed*; a triangulation is given by Fig. 1 if the heavily marked rectangle, whose sides are parallel to the coordinate axes, has sides of length a and b. In this figure a "point" of $\mathfrak{T}$ is marked by small circles.

EXAMPLE 7. The possible simultaneous *positions of two hands on a dial* are the points of a manifold; one describes a continuous curve thereon by moving the hands in any continuous and simultaneous fashion. Obviously this manifold is closed and equivalent to the torus.

EXAMPLE 8. The covering of a surface by a countable sequence of neighborhoods may be turned into an *abstract definition of the surface* by the following process: one starts with finitely or countably many copies E_1, E_2, ... of the unit disc in the xy-plane and identifies various points of these copies with each other in some suitable manner. Thus it is assumed that for each index pair i, j $(i \neq j)$ there is given an open subset E_{ij} of E_i and a pair of maps S_{ij} and S_{ji}, which are inverses of each other, such that S_{ij} (S_{ji}) is a neighborhood-true map of E_{ij} (E_{ji}) onto E_{ji} (E_{ij}). If p_i is a point of $E_{ij} \subset E_i$, then p_i is identified with the point p_j of $E_{ji} \subset E_j$, where p_j is the image of p_i under S_{ij}. Naturally we must assume that this identification relation is *transitive*, so that if p_i is identified with p_j and p_j with p_k, then the map S_{ik} carries p_i into p_k. If E_{ij} (and hence E_{ji}) is nonvoid, we speak of an overlap of E_i and E_j. The *connectedness* of the surface is guaranteed if it is impossible to split the E_i into two nonvoid classes such that two overlapping discs are always in the same class. Neighborhoods in the discs E_i are specified in the natural fashion. If $E_i^{(\nu)}$ ($\nu = 1, 2, \ldots$) denotes the compact subset of E_i defined by $x^2 + y^2 \leq (1 - 2^{-\nu})^2$, then it is clear that our two-dimensional manifold, which has been defined abstractly in the manner described, satisfies the axiom of countability; for the manifold is covered with countably many compact subsets $E_i^{(\nu)}$.

Example 9. As pointed out earlier, an *analytic form G* becomes a connected two-dimensional manifold if we regard the function elements of G as "points" and use as neighborhoods of such a point the analytic neighborhoods of § 2. It remains to be shown that this two-dimensional manifold satisfies the axiom of countability and that hence it may be called a surface.

Thus we must prove that G can be covered with countably many compact pieces. We take *first* as such pieces – each consisting of only *one* point – the irregular elements of G. They form a scattered set on G, and we have seen that there are only countably many of them. *Secondly* we add countably many pieces, consisting of regular elements, as follows. Let a be any rational point (that is, a value of z of the form $a_1 + ia_2$, where a_1 and a_2 are rational), let $Q(z - a)$ be a regular element, with center a, of G, and let n be the first of the natural numbers 0, 1, 2, ... such that 2^{-n} is less than the radius of convergence of Q. Then the inequality $|z - a| \leq 2^{-n}$ defines a compact part C of an analytic neighborhood of Q. Since the rational points a are countable, and since G contains at most countably many regular elements with center a, we obtain a countable set of pieces, C.

We must show now that any regular element of G is in one of these sets C. Let $\mathfrak{P}(z - c)$ be such an element, with center c, which converges in at least the circle $|z - c| < 2^{-m}$, where m is one of the numbers 0, 1, 2, ... Choose a rational point a whose distance from c is less than $\frac{1}{2} \cdot 2^{-m}$, and construct the associated $Q(z - a)$ by immediate analytic continuation of $\mathfrak{P}$. Its radius of convergence is greater than $\frac{1}{2} \cdot 2^{-m}$; hence the value n attached to Q, as described above, satisfies $n \leq m + 1$. The point c is interior to the circle with center a and radius 2^{-n}, for

$$|c - a| < 2^{-(m+1)} \leq 2^{-n}.$$

Thus inversely $\mathfrak{P}(z - c)$ is obtained from the element $Q(z - a)$ of G by immediate analytic continuation; thus $\mathfrak{P}$ is a "point" of the set C determined by Q.

§ 6. Specialization; in particular, differentiable and Riemann surfaces

If $\mathfrak{e}$ is an element of G with the representation

$$z = P(t), \qquad u = Q(t)$$

valid for $|t| < r \ (r > 0)$, then for each t, $|t| < r$, there is an element $\mathfrak{e}_t$,

$$z = P(t + t'), \qquad u = Q(t + t'),$$

which is obtained by rearranging the development of $\mathfrak{e}$ in powers of t'. This transformation $t \to \mathfrak{e}_t$ is a topological map of the disc $|t| < r$ in the t-plane onto a certain neighborhood of the point $\mathfrak{e}$ on G: thus the parameter t appears as a continuous function defined in this neighborhood on G. We call t a *local uniformizing parameter* at the point $\mathfrak{e}$. Every other local uniformizing parameter τ at $\mathfrak{e}$ can be represented in the form

$$\tau = \gamma_1 t + \gamma_2 t^2 + \cdots \qquad (\gamma_1 \neq 0)$$

for sufficiently small t. A complex valued function f, defined in a domain of G, will be called *regular analytic* at the point $\mathfrak{e}$ of this domain if it can be expressed, in some neighborhood of $\mathfrak{e}$, by a regular power series in the local uniformizing parameter t at $\mathfrak{e}$:

$$f = a_0 + a_1 t + a_2 t^2 + \cdots.$$

It is immaterial which local parameter t at $\mathfrak{e}$ is used; if such a representation is possible with *one* local parameter, then f has the same type of representation in terms of any other local parameter. Naturally, the representation is valid only in some neighborhood of $\mathfrak{e}$ which depends on the choice of the local parameter.

If the development of z $[z = P(t)]$ in the representation of the function element $\mathfrak{e}$ contains no negative powers of t, then the constant initial term z_0 in this development depends only on $\mathfrak{e}$ and not on the representation of $\mathfrak{e}$; z_0 is called the *value* of the complex variable z at the "point" $\mathfrak{e}$. If the development of z begins with negative powers of t, then that is the case for every representation of the function element $\mathfrak{e}$, and the value of z at the point $\mathfrak{e}$ is $= \infty$. *Thus z is seen to be a uniform (single-valued) function, defined on the "surface" G, which is regular analytic except at isolated points of G where z becomes infinite.* Similarly, we may regard u as a uniform function on G, regular analytic except for poles. The independent argument in each of these functions is not a complex variable in the usual sense (that is, not a point varying in a plane domain), but rather a variable point on the "Riemann surface" G.

For the statement that z and u are analytic functions on the surface G, it is essential that *G is given not merely as a surface in the sense of analysis situs.* One can talk of a continuous function on a surface for which only analysis-situs properties are considered; but one cannot talk of "continuously differentiable," "analytic" (or even "entire rational"), and such like functions. In order to pursue analytic function theory on a surface $\mathfrak{F}$, in a fashion analogous to function theory in the plane, an explanation (besides the definition of the surface) must be given to specify the meaning of the term "analytic function on the surface"; this specification must be such that all theorems which are valid "in the small" for analytic functions in the plane carry over to this more general situation. Theorems valid "in the small" are those which affirm a statement about a certain neighborhood of a point without making any statement about the size of that neighborhood. As done above, the functions analytic at a point may be introduced as power series in a local parameter at the point. If, from among the possible topological coordinate systems (x, y) for a neighborhood of a point, we single out those for which the combination $t = x + iy$ serves as a local parameter, then the surface becomes a Riemann surface. This formulation of the concept of a Riemann surface, first developed in intuitive form by F. Klein in his monograph *Über Riemann's Theorie der algebraischen Funktionen und ihrer Integrale,*[14] is more general than the formulation which Riemann himself used in his fundamental work on the theory of analytic functions. There can be no doubt but that the full simplicity and power of Riemann's ideas become apparent only with this general formulation. Moreover, Riemann himself

---

[14]) Leipzig 1882. See also Klein, *Neue Beiträge zur Riemannschen Funktionentheorie*, Math. Ann, **21** (1883), §§1–3 [pp. 146–151]. Surfaces, closed by boundary identifications, as carriers of analytic functions appear earlier: Riemann, Art. 12 of *Theorie der Abelschen Funktionen, Werke,* p. 121; H. A. Schwarz in his fundamental paper of 1870 on the integration of the partial differential equation $\partial^2u/\partial x^2 + \partial^2u/\partial y^2 = 0$, *Gesammelte mathematische Abhandlungen,* II, 161; Dedekind, *Jour. f. Math.,* **83** (1877) 274 ff. Surfaces imbedded in space were first used, though only for investigations in analysis situs, by Tonelli (1875, Atti dei Lincei, ser II, v. **2**) and Clifford (1876, *Mathematical Papers,* 249 ff). Klein himself, as he relates in the preface to his monograph, *Über Riemann's Theorie* (p. IV), derived the initial idea for his formulation from a chance oral remark of Prym (1874). Klein considers only closed surfaces. The most general concept is probably found explicitly first in Koebe's work; see, for example, Göttinger Nachrichten (1908), 338–339, footnote.

laid the foundations for this general formulation by developing the concept of an *n*-dimensional manifold in his Habilitationsvortrag.[15] It may be assumed with certainty that for Riemann the ideas developed in that lecture were closely related to his investigations in function theory; however, he did not mention this relation explicitly.

After these indications, we will now describe in precise fashion how the concept of a surface may be specialized to a "differentiable" or "Riemann" surface. This specialization depends upon *local* coordinates and the restriction of admissible local coordinates by means of a *group* of coordinate transformations. We start with a general preparatory remark.

Let $\mathfrak{F}$, $\mathfrak{F}^*$ be two manifolds, $\mathfrak{p}_0$ a point of $\mathfrak{F}$, $\mathfrak{p}_0^*$ a point of $\mathfrak{F}^*$, $\mathfrak{G}$ an open subset of $\mathfrak{F}$ containing $\mathfrak{p}_0$, and $\mathfrak{G}^*$ an open subset of $\mathfrak{F}^*$ containing $\mathfrak{p}_0^*$. Let S (with inverse S^{-1}) be a topological map, $\mathfrak{p} \to \mathfrak{p}^*$ ($\mathfrak{p}^* \to \mathfrak{p}$), of $\mathfrak{G}$ onto $\mathfrak{G}^*$ which carries $\mathfrak{p}_0$ into $\mathfrak{p}_0^*$. Our only interest in this map is *local*; by that we mean the following. Let $\mathfrak{G}_1$ ($\mathfrak{G}_1^*$) be an open subset of $\mathfrak{F}$ ($\mathfrak{F}^*$) containing $\mathfrak{p}_0$ ($\mathfrak{p}_0^*$), and let S_1 be a topological map of $\mathfrak{G}_1$ onto $\mathfrak{G}_1^*$ which maps $\mathfrak{p}_0$ into $\mathfrak{p}_0^*$. Then S and S_1 are *locally not distinct at* $\mathfrak{p}_0$ if there exists a neighborhood $\mathfrak{U}$ of $\mathfrak{p}_0$ contained in both $\mathfrak{G}$ and $\mathfrak{G}_1$ such that S and S_1 agree on $\mathfrak{U}$. (Then, of course, there exists a neighborhood of $\mathfrak{p}_0^*$, $\mathfrak{U}^*$ contained in both $\mathfrak{G}^*$ and $\mathfrak{G}_1^*$ in which S^{-1} and S_1^{-1} agree.) In this sense we speak of a *local topological map* between $\mathfrak{F}$ and $\mathfrak{F}^*$ with center ($\mathfrak{p}_0$, $\mathfrak{p}_0^*$). It is clear how one can carry out the *composition* of such local maps. Let $\mathfrak{p}_0$, $\mathfrak{p}_0^*$, $\mathfrak{p}_0^{**}$ be points of the manifolds $\mathfrak{F}$, $\mathfrak{F}^*$, $\mathfrak{F}^{**}$; let S be a topological map of the open set $\mathfrak{G}$, containing $\mathfrak{p}_0$, onto the open set $\mathfrak{G}^*$, such that the image of $\mathfrak{p}_0$ is $\mathfrak{p}_0^*$; let S^* be a topological map of the open set $\mathfrak{G}_1^*$, containing $\mathfrak{p}_0^*$, onto the open set $\mathfrak{G}^{**}$, such that the image of $\mathfrak{p}_0^*$ is $\mathfrak{p}_0^{**}$. Shrink $\mathfrak{G}$ to $\mathfrak{g}$, the image under S^{-1} of the intersection of $\mathfrak{G}^*$ and $\mathfrak{G}_1^*$. The map SS^* is defined at the points of $\mathfrak{g}$ and maps $\mathfrak{g}$ topologically onto an open set $\mathfrak{g}^{**}$ on $\mathfrak{F}^{**}$. If one replaces S and S^* by maps which are locally not distinct from S and S^*, then SS^* is replaced by a map which is locally identical with SS^*. Thus we can speak of the composition of local topological maps.

A local coordinate system (x, y) at the point $\mathfrak{p}_0$ of the surface $\mathfrak{F}$ is defined by a topological map S of an open subset $\mathfrak{G}$ of $\mathfrak{F}$, which contains $\mathfrak{p}_0$, onto an open set $\mathfrak{E}$ of the xy-plane, such that $\mathfrak{p}_0$ maps into the origin $O = (0, 0)$. In this process the local point of view, described above, at the center ($\mathfrak{p}_0$, O) becomes significant: the replacement of S by a locally identical map at this center replaces (x, y) by a locally identical coordinate system. The transfor-

---

[15]) *Über die Hypothesen, welche der Geometrie zugrunde liegen, Werke,* 2nd edition, 272–287; edited with a commentary by H. Weyl, 3rd edition, Berlin 1923.

mation from one local coordinate system (x, y) at the point $\mathfrak{p}_0$ to another (x^*, y^*) amounts to a local topological map $(x, y) \rightleftarrows (x^*, y^*)$ with center (O, O).

On the basis of the composition explained above, one can consider *groups* Γ of local topological coordinate transformations which leave the center O fixed. For example, one can put in Γ those coordinate transformations

$$x^* = \phi(x, y), \qquad y^* = \psi(x, y)$$

which leave the origin fixed and are such that in some neighborhood of the origin (a disc with center O) the functions ϕ and ψ are defined and continuously differentiable, and also the functional determinant

$$J = \frac{\partial \phi}{\partial x}\frac{\partial \psi}{\partial y} - \frac{\partial \phi}{\partial y}\frac{\partial \psi}{\partial x}$$

does not vanish at the origin. Such a "*differentiable*" *map* has an inverse of the same sort; the map is locally topological at O and these differentiable maps form a group Γ. As a consequence of the local point of view, the membership of a coordinate transformation in Γ is not affected by shrinking the neighborhood of the origin in which it is defined. Also, the continuously differentiable transformations in a neighborhood of O, for which the functional determinant J is *positive*, constitute a group Γ^+, which is clearly a subgroup (of index 2) of Γ. We call its elements the *positive differentiable maps*. If, instead of the two real coordinates x and y, we use one complex coordinate, $t = x + iy$, then the transformation given by

$$t^* = a_1 t + a_2 t^2 + \cdots, \tag{6.1}$$

where $a_1 \neq 0$ and the power series converges in some disc $|t| < r$ of positive radius r, is called *conformal* at the point $t = 0$. The locally conformal transformations of t at O form a group Γ_c. If one expresses the transformation (6.1) in terms of the real and imaginary parts, x and y of t, and x^* and y^* of t^*, the functional determinant is $|a_1|^2$. Hence Γ_c is a subgroup of the group Γ^+ of positive differentiable transformations.

A special class of surfaces is now determined by a given group Γ of local coordinate transformations at the origin as follows. A class of local coordinate systems at the point $\mathfrak{p}_0$ of the surface is called *admissible* if any two are related by a transformation of the local group Γ ("*surface of type* Γ"). The best procedure is the following: for an arbitrary point $\mathfrak{p}_0$ choose *one* topological map of an open subset of the surface $\mathfrak{F}$, containing $\mathfrak{p}_0$, onto an open set in the xy-plane such that $\mathfrak{p}_0$ maps into the origin. The admissible coordinate systems at $\mathfrak{p}_0$ are those which can be obtained from this one by means of a local transformation (at O) of the group Γ. Clearly this includes the demand that if one coordinate system at $\mathfrak{p}_0$ is admissible, then any other locally identical with that at $(\mathfrak{p}_0, O)$ is admissible. Furthermore, the following essential condition must be satisfied: if (x, y) is an admissible local coordinate system at $\mathfrak{p}_0$, then there is an *associated neighborhood* $\mathfrak{U}_\varepsilon$, $x^2 + y^2 < \varepsilon^2$, of the origin, such that $x - x_1, y - y_1$ is an admissible local coordinate system at the point $\mathfrak{p}_1$ corresponding to (x_1, y_1) *provided that* (x_1, y_1) *is a point of the neighborhood* $\mathfrak{U}_\varepsilon$. That is, the coordinate system $x - x_1, y - y_1$ arises from the chosen coordinate system at $\mathfrak{p}_1$ via a transformation of Γ. [In calling the preimages of such $\mathfrak{U}_\varepsilon$ (neighborhoods of the origin in admissible local coordinate systems) neighborhoods of $\mathfrak{p}_0$ on $\mathfrak{F}$, we are replacing the neighborhoods by means of which $\mathfrak{F}$ was defined as a continuous manifold by an equivalent set of neighborhoods. Thus we need not worry over calling $\mathfrak{U}_\varepsilon$ a neighborhood of $\mathfrak{p}_0$ on $\mathfrak{F}$.] In this matter it is important that if the functional determinant of a continuously differentiable coordinate transformation is nonzero (positive) at the origin, then it possesses the same property in some neighborhood of the origin.

The surfaces of type Γ = group of local differentiable maps, Γ^+ = group of positive local differentiable maps, and Γ_c = group of local conformal maps, are called *smooth*, *smooth oriented*, or *Riemann surfaces*, respectively. Thus every Riemann surface is a smooth oriented surface. A local conformal parameter $t = x + iy$, in a neighborhood of $\mathfrak{p}_0$ on a Riemann surface, will be called a *local (uniformizing) parameter at* $\mathfrak{p}_0$. Every subdomain of a surface of type Γ is a surface of type Γ, whatever the group Γ may be. A function f defined in a neighborhood of a point $\mathfrak{p}_0$ on the smooth surface $\mathfrak{F}$ is called continuously differentiable at $\mathfrak{p}_0$ if it may be expressed in terms of some admissible local coordinate system x, y at $\mathfrak{p}_0$ as a function $f(x, y)$ which is continuously differentiable in some neighborhood of the origin. This concept is obviously independent of the choice of the admissible local coordinates. The function is continuously differentiable, not only at $\mathfrak{p}_0$, but also in some neighborhood of $\mathfrak{p}_0$. Similarly, a (complex-valued) function defined in a neighborhood of a point $\mathfrak{p}_0$ on a Riemann surface is *regular analytic* at $\mathfrak{p}_0$ if it can be expressed as a power series

$$f(t) = A_0 + A_1 t + A_2 t^2 + \cdots$$

in terms of a local parameter t at $\mathfrak{p}_0$ which converges in some neighborhood $|t| < r$ of the origin. The concept of a regular analytic function is conformally invariant; that is, it is independent of the choice of local parameter.

Our discussion results in the following scale of successive specializations: (topological) surface → smooth surface → smooth oriented surface → Riemann surface. If a concept applicable on a Riemann surface has validity (is invariant) on arbitrary smooth or even arbitrary topological surfaces, then, for the sake of generality, one wishes to introduce it in that wider context. For example, the concept of an ellipse is invariant under all collineations, and one would rather associate the concept with projective geometry than with the more special metric-Euclidean geometry. But, of course, one is not forced to take this approach. That the lowest step is labeled with the particular name *topology* is actually somewhat arbitrary; it is justified only by the expansion of this discipline in recent decades.

With the help of the *neighborhood concept* and its axiomatics we have been able to characterize the topological (or "continuous") manifolds by their intrinsic properties. To date it has not been possible to treat the smooth (or Riemannian) manifolds in a similar fashion. Here we remain forced to determine the class of admissible, or equally justified, local coordinate systems (related by the transformations of a definite group Γ) by a choice of *one* of them. The disadvantage of differential geometry, as compared with Euclidean or projective geometry and also topology, is that we are not in a position to found it on invariant basic concepts (fundamental relations) and axioms therefor. The situation is no different for conformal geometry on a Riemann surface.

That the unit sphere in three-dimensional Euclidean space is a smooth oriented surface may be seen as follows. For an arbitrary point A on the sphere, introduce new Cartesian coordinates (x_1, x_2, x_3) by an orthogonal transformation of (ξ_1, ξ_2, ξ_3) so that $A = (0, 0, 1)$; call (x_1, x_2) an admissible local coordinate system at the "north pole" A (it is usable on the whole northern hemisphere $x_3 > 0$). The complex z-plane, without the point at infinity, becomes a Riemann surface if we take at each point a the difference $z - a$ as a local parameter. Also the surface obtained by including the point at infinity, which, on the basis of the familiar stereographic projection, we call the z-sphere, is a Riemann surface if $1/z$ is used for a local parameter at the point at infinity.

We list now several invariant concepts concerning the behavior of functions and curves on Riemann surfaces.

If $f(\mathfrak{p})$ is regular at $\mathfrak{p}_0$, if t is a local parameter at $\mathfrak{p}_0$, and if the development of $f(\mathfrak{p})$ in powers of t begins with the term $a_m t^m$ ($a_m \neq 0$), then $\mathfrak{p}_0$ is a *zero of order m of f*. We say also that f has *order m* at the point $\mathfrak{p}_0$. If $\tau(\mathfrak{p})$ is any other local parameter at $\mathfrak{p}_0$,

$$t = c_1\tau + c_2\tau^2 + \cdots \qquad (c_1 \neq 0),$$

then the development of f in powers of τ begins with the term $a_m c_1^m \tau^m$; the "invariance" of the order m of a zero follows. If in some neighborhood of the point $\mathfrak{p}_0$, except at $\mathfrak{p}_0$ itself, the function f has a development

$$f = \frac{a_{-n}}{t^n} + \cdots + \frac{a_{-1}}{t} + a_0 + a_1 t + \cdots \qquad (a_{-n} \neq 0, n \text{ a positive integer}),$$

then f has a *pole of order n* at $\mathfrak{p}_0$. We also say that f has *order* $-n$ at $\mathfrak{p}_0$. The proof of invariance is as above. If $f(\mathfrak{p})$ has order k (which may be positive, zero, or negative) at $\mathfrak{p}_0$ and $g(\mathfrak{p})$ has order l, then $f \cdot g$ has order $k + l$ and f/g has order $k - l$ at $\mathfrak{p}_0$.

If a function f is uniform and regular analytic in a neighborhood of $\mathfrak{p}_0$, except at $\mathfrak{p}_0$ itself, then either one can assign a value to f at $\mathfrak{p}_0$ such that f becomes regular at $\mathfrak{p}_0$, or f has a pole at $\mathfrak{p}_0$, or f has an *essential singularity* at $\mathfrak{p}_0$. In the last case, $f(\mathfrak{p})$ comes arbitrarily close to every value in any neighborhood of $\mathfrak{p}_0$.

If, z, u are two functions, regular except for poles in a domain $\mathfrak{G}$ of a Riemann surface, then each rational expression $R(z, u)$ in z and u is regular in $\mathfrak{G}$ except for poles. A rational expression $R(z, u)$ in two independent variables z, u is of the form $F(z, u)/G(z, u)$, where F and G are polynomials in these variables. Our claim is obvious if R is a polynomial F: replace z and u in $F(z, u)$ by their developments in powers of a local parameter at a point of the domain. The corresponding result follows for the quotient F/G, since the quotient of the two power series $f(t)/g(t)$ may be written as a power series (with only a finite number of negative powers of t). But we must rule out the case where $g(t)$ is identically zero. By analytic continuation in the connected domain $\mathfrak{G}$, this case arises only if $G(z, u)$ becomes identically zero when z and u are replaced by the analytic functions denoted by the same letters. This case must indeed be ruled out. But, with this case excluded, even if a direct substitution of values of z and u yields an indeterminate form $0/0$, there is in fact no point of indetermination.

We discussed above the sense in which an analytic form may be regarded as a Riemann surface. But the concepts "analytic form" and "Riemann surface" are not identical. With an analytic form (z, u) we are given not merely a Riemann surface, but at the same time two functions z and u on the surface, regular except for poles. Also, z and u satisfy the following condition: there is no pair of distinct points $\mathfrak{p}_1^0$ and $\mathfrak{p}_2^0$, with local parameters t_1 and t_2, and a pair of power series $P(t)$ and $Q(t)$ such that

$$\begin{aligned} z &= P(t_1), \qquad u = Q(t_1) \qquad \text{in a neighborhood of } \mathfrak{p}_1^0, \\ z &= P(t_2), \qquad u = Q(t_2) \qquad \text{in a neighborhood of } \mathfrak{p}_2^0. \end{aligned}$$

An arbitrary Riemann surface becomes an analytic form if we single out two functions z, u on it, regular except for poles, which satisfy the condition just formulated (and such that z does not reduce to a constant). If there exists *one* such pair of functions z, u, then such pairs may be chosen in infinitely many different ways; for example, instead of z and u, I can use any two linear combinations of z and u:

$$z' = az + bu, \qquad u' = Az + Bu$$
$$(a, b; A, B \text{ constant}; \quad aB - bA \neq 0).$$

That *there actually is a function pair* (z, u), *that is, an analytic form, associated with any given Riemann surface,* is a fundamental fact of Riemannian function theory; with the aid of the Thomson-Dirichlet principle, which Riemann used for the same purpose, the proof for closed Riemann surfaces will be given in Chapter II of this book.

A real-valued function U is called a *harmonic*, or a *potential, function* at the point $\mathfrak{p}_0$ if there exists a function which is regular analytic at $\mathfrak{p}_0$ whose real part coincides with U in some neighborhood of $\mathfrak{p}_0$. If U is harmonic at every point of a domain, and not a constant, then U has neither a maximum nor a minimum at any point of the domain. This follows in the familiar fashion from the fact that if $U(t)$ is harmonic in a neighborhood of $t = 0$, then $U(0)$ is the mean value of U on small circles in the t-plane with center at the origin. By the principle of analytic continuation it follows that if U is constant in a neighborhood of some point of its domain of definition, then U must be constant in the whole (connected!) domain. Therefore, *except for constants, there is no (uniform) function which is harmonic at every point of a closed Riemann surface; a fortiori, there is no (uniform) function which is regular analytic at every point of the surface.*

The sense in which one can speak of once, twice, thrice, ... continuously differentiable functions on a Riemann surface is clear. A curve $\mathfrak{p} = \mathfrak{p}(\lambda)$ $(0 \leq \lambda \leq 1)$ may be represented, in a neighborhood of any point $\mathfrak{p}(\lambda_0) = \mathfrak{p}_0$ of the curve, in the form $t = t(\lambda)$, where t is a local parameter at $\mathfrak{p}_0$ which maps a neighborhood of $\mathfrak{p}_0$ one-to-one and conformally onto a domain of the t-plane. If, for real λ sufficiently close to λ_0 and in the interval $[0, 1]$, $t(\lambda)$ may be expressed as a convergent power series $b_1(\lambda - \lambda_0) + b_2(\lambda - \lambda_0)^2 + \cdots$, with $b_1 = (dt/d\lambda)_{\lambda = \lambda_0} \neq 0$, then the curve is called *analytic* at $\lambda = \lambda_0$. The proof of invariance is trivial. By using the local parameter τ defined by $t = b_1\tau + b_2\tau^2 + \cdots$, an arc of the curve containing $\mathfrak{p}(\lambda_0)$ appears as an interval of the real axis in the τ-plane. An *analytic curve* is to be understood to mean a curve which is analytic at each value of the parameter in [0,1].

If we know only that the derivative $dt/d\lambda$ exists and is continuous in an interval about λ_0 and $(dt/d\lambda)_{\lambda=\lambda_0} \neq 0$, then we say that the *curve is continuously differentiable at* λ_0. Let

$$\mathfrak{p} = \mathfrak{p}^1(\lambda) \quad \text{and} \quad \mathfrak{p} = \mathfrak{p}^2(\lambda) \tag{6.2}$$

be two curves, starting at the same point $\mathfrak{p}_0 = \mathfrak{p}^1(0) = \mathfrak{p}^2(0)$, which are continuously differentiable at $\lambda = 0$. Let

$$t = t^1(\lambda), \qquad t = t^2(\lambda) \qquad (0 \leq \lambda \leq \lambda_1)$$

be the images of initial arcs of the two curves in the t-plane, where t is a local parameter at $\mathfrak{p}_0$. Let θ be the angle, determined to within an additive integral multiple of 2π, between these two images at the origin in the t-plane; θ is determined by

$$\left(\frac{dt^1}{d\lambda}\right)_{\lambda=0} = r_1 e^{i\theta_1}, \qquad \left(\frac{dt^2}{d\lambda}\right)_{\lambda=0} = r_2 e^{i\theta_2} \qquad (r_1, r_2 > 0;\ \theta_1, \theta_2 \text{ real}),$$
$$\theta = \theta_2 - \theta_1 .$$

Then we shall call θ the *angle* between the two curves (6.2) on the Riemann surface at $\mathfrak{p}_0$. This angular measure is invariant, since the change from one local parameter t to another τ is accomplished by an isogonal map of an open set containing the origin in the t-plane onto an open set of the τ-plane. Thus *there exists an invariant angular measure on a Riemann surface.*[16] With the introduction of this angular measure, the concept of a "conformal" map, explained above, coincides with the concept of an isogonal map.

If $\mathfrak{F}$ and $\mathfrak{F}^*$ are any two Riemann surfaces, then a one-to-one map $\mathfrak{F} \rightleftarrows \mathfrak{F}^*$ is called *conformal* if it is everywhere locally conformal. Two Riemann surfaces which can be mapped conformally onto each other are *(conformally) equivalent* and are to be regarded as different representations of one and the same ideal Riemann surface. The intrinsic properties of a Riemann surface will include only those properties which are invariant under conformal maps; that is, those properties which, if possessed by *one* Riemann surface $\mathfrak{F}$, are possessed by every *equivalent* surface. Obviously all topological properties are intrinsic properties of a Riemann surface; similarly with those properties belonging to the surface by virtue of its *smoothness.*

A surface, free of singularities, in Euclidean space, such as a sphere or torus, on which (after stipulation of a sense of rotation) there exists a natural angular measure may be regarded, in a unique way, as a Riemann surface; namely, the angular measure on the Riemann surface is to coincide with the

---

[16]) That there exists, in a certain sense, an invariant measure of length, is a deep fact in uniformization theory. See § 21 of this book.

natural angular measure. For this purpose one must show that every point of the surface has a neighborhood which may be mapped isogonally (in the Euclidean sense) onto a plane domain.[17]

For the sphere, stereographic projection yields a function z, regular on the sphere except for a single pole of first order, which assumes each value, including ∞, exactly once. Any other function regular except for poles on the sphere is a rational function of z; thus the theory of functions regular except for poles on the sphere is in effect the same as the theory of rational functions of one variable z. To be sure, the choice of the independent variable z is to some extent arbitrary; besides z, each function z' obtained by a linear transformation on z will serve:

$$z' = \frac{az + b}{cz + d} \qquad (a, b, c, d \text{ constants}, ad - bc \neq 0).$$

The position and multiplicity of the zeros and poles of a function, analytic on the sphere and without essential singularities, may be stipulated arbitrarily, provided only that the total order of the zeros equals the total order of the poles.

The situation on the torus is quite different (see Chapter II, § 18). As the intrinsic reason for this difference in behavior of functions on the torus and functions on the sphere we can almost always point to the following fact (which falls in the domain, not of function theory, but of analysis situs): on the torus there exists a pair of closed curves ($\phi = 0$ and $\psi = 0$ in the notation of § 5, Example 6), crossing at one point, which does not separate the torus; such a pair does not exist on the sphere. If we map the torus conformally onto the point-lattice manifold $\mathfrak{T}$ (§ 5), then the functions on the torus, without essential singularities, appear as uniform functions of the complex variable $w = \phi + i\psi$ which are regular except for poles and *doubly periodic* with periods 2π and $2\pi\, ir\,(R^2 - r^2)^{-\frac{1}{2}}$; that is, as *elliptic functions* [with purely imaginary period ratio $= ir\,(R^2 - r^2)^{-\frac{1}{2}}$].

Two tori are always equivalent in the sense of analysis situs. But in general they cannot be mapped conformally onto each other; rather, that is possible if and only if the value of the "module" $rR^{-2}(R^2 - r^2)^{\frac{1}{2}}$ is the same for the two tori. Here the term "module" has the following meaning: if we are given a family of Riemann surfaces, any two of which are equivalent in the sense

---

[17]) See L. Lichtenstein, *Beweis des Satzes, dass jedes hinreichend kleine, im wesentlichen stetig gekrümmte, singularitätenfreie Flächenstück auf einem Teil einer Ebene zusammenhängend und in den kleinsten Teilen ähnlich abgebildet werden kann,* Abhandlungen der Preussischen Akademie der Wissenschaften vom Jahre 1911, Anhang.

of analysis situs, and if there is a number associated with each surface such that any two *conformally* equivalent surfaces possess the same number, then that number, regarded as a functional on the family, is a module of the family. The fact that, in general, equivalence in the sense of analysis situs does not entail the conformal equivalence of Riemann surfaces is of fundamental importance. Let the points of one torus be represented by the point-lattices in the w_1-plane with periods 2π and $2\pi\, ia_1 = 2\pi\, ir_1\, (R_1^2 - r_1^2)^{-\frac{1}{2}}$; let the points of the second torus be the point-lattices in the w_2-plane with periods 2π and $2\pi\, ia_2$ $(a_2 > 0)$. If there is a conformal map of one torus onto the other, then to each w_1-lattice there corresponds a w_2-lattice such that if the lattice w_1 is subjected to an infinitesimal displacement dw_1, then the image lattice w_2 undergoes an infinitesimal displacement dw_2 such that the ratio dw_2/dw_1 depends only on the lattice w_1 and not on the *direction* of the displacement dw_1. Regarded as a function of the lattice w_1, this ratio is a regular analytic function on the first torus, and therefore must be a constant, A. It follows that the conformal map must be given by the formula

$$w_2 = Aw_1 + B \qquad (A, B \text{ constants}) \tag{6.3}$$

in the following sense: when one substitutes for w_1 in (6.3) all the points of a w_1-lattice, one obtains all the points in the corresponding w_2-lattice. A simple lattice argument of a number theoretic nature shows that this is possible only when $A = \pm 1$ or $= \pm i/a_1$; in the first case $a_2 = a_1$ and in the second case $a_2 = 1/a_1$. In either case the equation

$$a_1 + \frac{1}{a_1} = a_2 + \frac{1}{a_2}$$

is a necessary (and clearly, also sufficient) condition that the two tori be conformally equivalent. This agrees with our claim, since

$$\frac{r}{\sqrt{R^2 - r^2}} + \frac{\sqrt{R^2 - r^2}}{r} = \frac{R^2}{r\sqrt{R^2 - r^2}}.$$

We close with a few general remarks on the *concept of a Riemann surface.* The basic idea responsible for their introduction is by no means restricted to complex function theory. A function of two real variables x, y is a *function in the plane*; but it makes just as much sense to investigate functions on the sphere, the torus, or any surface. To be sure, so long as one is concerned only with the behavior of functions "in the small" – and that is what most investigations in analysis are concerned with – then the concept of a function of two real variables is general enough; for a neighborhood of any point on a two-dimensional manifold has a representation in terms of x, y (or $x + iy$).

But as soon as one proceeds to investigate the behavior of functions "*in the large*," the functions in the plane form an important but *special case among infinitely many others with equal claims*. Riemann and Klein taught us not to stop at this special case. As applied to complex function theory this means *before one proceeds with the study of any class of functions, the surface which is the domain of the independent argument must be defined; then it must be stipulated what "analytic function" on this surface shall mean, so that the surface becomes a Riemann surface. Only then can one get busy with the functions themselves*. Correspondingly, in investigating analytic functions, one proceeds in two steps: first one attempts to characterize the Riemann surface on which each function lives by intrinsic properties. In doing this one can follow a ranking according to whether the properties in question are invariant under topological maps, or only under differentiable maps, or (as in the case of a module) only under conformal maps. Only at the second stage do those properties in which functions on the same Riemann surface differ (such as the position and order of zeros and poles) come under consideration. From this point of view, Weierstrass' concept of an analytic form plays only a secondary role: it arises only when one combines two functions on one surface. Instead of two, it is a natural step to consider a set of three or four or more functions on one Riemann surface. Geometrically speaking, this step amounts to going from the study of plane analytic "curves" to that of curves in three or four more dimensional space.

The uniform functions, regular except for poles on a Riemann surface $\mathfrak{F}$, will be called *meromorphic functions* or "*functions*" *on the surface*. For *closed* Riemann surfaces we will obtain a view of all these functions in Chapter II. One can also consider the following more general class of functions attached to $\mathfrak{F}$. Start with a function element on $\mathfrak{F}$; that is, a power series in integral powers of some local parameter t at a point $\mathfrak{p}_0$ of $\mathfrak{F}$. One can then attempt to continue this function element analytically along all possible paths on $\mathfrak{F}$, in a fashion analogous to that described for the case of the plane in § 1. Suppose that there exists a unique continuation along every path; that is, one meets at worst poles, but never points through which continuation is impossible ("natural boundaries") or points at which the continuation becomes multiple valued (branching relative to $\mathfrak{F}$). As is shown by the example $w = \phi + i\psi$ on the torus, the resulting function need not by any means be uniform. Rather, such a function derived by analytic continuation will be uniform on a certain covering surface spread out over $\mathfrak{F}$ without boundaries and without branching. In many questions, particularly in *uniformization theory*, it is of great importance to expand the class of uniform functions on $\mathfrak{F}$ to the larger class of all (finitely or infinitely many-valued) functions which are unbranched and without essential singularities or natural boundaries on $\mathfrak{F}$. This places on us the duty of considering in this chapter, the remainder of which is devoted to the topological questions which are fundamental to function theory, the *covering surfaces* associated with each surface $\mathfrak{F}$.

§ 7. Orientation

An admissible local coordinate system x, y at the point $\mathfrak{p}_0$ of the smooth manifold $\mathfrak{F}$ determines a unique *sense of rotation* at $\mathfrak{p}_0$ by means of the following agreement. Another such coordinate system x^*, y^* determines the same sense of rotation at $\mathfrak{p}_0$ if and only if the functional determinant J of the transformation $(x, y) \to (x^*, y^*)$ is positive at the origin. Thus two opposite senses of rotation are possible at a point $\mathfrak{p}_0$. On a smooth *oriented* surface there is specified a unique *positive* sense of rotation at each point, given by any admissible local coordinate system at the point.

The power of differential calculus is that it *linearizes* all problems by going back to the "infinitesimally small," but this process can be used only on smooth manifolds. Thus our distinction between the two senses of rotation on a smooth manifold rests on the fact that a continuously differentiable coordinate transformation leaving the origin fixed can be approximated by a linear transformation at O and one separates the (nondegenerate) homogeneous linear transformations into positive and negative according to the sign of their determinants. Also the *invariance of the dimension* for a *smooth* manifold follows simply from the fact that a linear substitution which has an inverse preserves the number of variables. The proof of the invariance of the dimension for *topological* manifolds is far from being so simple. We do not need to worry about this, but we shall consider the corresponding problem for orientation: to extend the concept of orientation to arbitrary topological surfaces proves to be essentially simpler.

Consider a point A in the xy-plane and a closed curve $\mathfrak{C}$ not passing through A. While the variable point P traces $\mathfrak{C}$ once in the prescribed sense, we follow the continuous change of the angle ϕ between the ray AP and the positive direction of the x-axis. The increase of ϕ, that is, the difference between the value of ϕ at the end and the value of ϕ at the beginning, is an integer n, provided we use that angular measure for which 1, not 2π, is a complete revolution. The number n is called *the order of A relative to* $\mathfrak{C}$, in symbols

$$n = \operatorname*{ord}_{\mathfrak{C}}(A);$$

we say that the curve $\mathfrak{C}$ goes around A n times. If A and B are two points of the plane which can be joined by a curve σ which does not meet $\mathfrak{C}$, then A and B have the same order relative to $\mathfrak{C}$. This follows by letting a point wander along σ from A to B and observing that the order of that point relative to $\mathfrak{C}$ varies continuously; since its value is always an integer, it must be constant.

The topological distinction between two senses of rotation at a point $\mathfrak{p}_0$ of an arbitrary two-dimensional manifold $\mathfrak{F}$ rests on the following theorem.

Let there be given two open sets $\mathfrak{G}$, $\mathfrak{G}'$ in the xy-plane and a topological map S of $\mathfrak{G}$ onto $\mathfrak{G}'$. Also let $\mathfrak{E}$ be a circular disc contained in $\mathfrak{G}$. Then there exists a definite sign $\delta = +1$ or -1 such that if A is a point of $\mathfrak{E}$ and $\mathfrak{C}$ is a closed curve in $\mathfrak{E}$ not through the point A, then n, the order of A relative to $\mathfrak{C}$, and n', the order of the image point A' relative to the image curve $\mathfrak{C}'$, are related by $n' = \delta n$.

To begin with we prove less, namely that δ is an integer independent of $\mathfrak{C}$. Then it follows easily first that $\delta = \delta_A$ must have one of the values ± 1, and second that δ_A is independent (not only of $\mathfrak{C}$ but also) of A. First, among the closed curves $\mathfrak{C}$ in $\mathfrak{E}$ which do not pass through A are found all those whose image $\mathfrak{C}'$ is a sufficiently small circle with center at A'. Since A' has the order $n' = 1$ relative to such a circle, δ must be a divisor of 1 and hence $+1$ or -1. Second, compare the point A in $\mathfrak{E}$ with the center O of $\mathfrak{E}$ and take a circle $\mathfrak{k}$ in $\mathfrak{E}$, with center O, which encloses A. Join O with A inside $\mathfrak{E}$ with the segment $\sigma = OA$. Then

$$\operatorname*{ord}_{\mathfrak{k}}(O) = 1, \qquad \operatorname*{ord}_{\mathfrak{k}'}(O') = \delta_O, \qquad \operatorname*{ord}_{\mathfrak{k}}(A) = 1, \qquad \operatorname*{ord}_{\mathfrak{k}'}(A') = \delta_A.$$

But since O' is joined to A' by a curve σ' which does not meet $\mathfrak{k}'$, O' and A' must have the same order relative to $\mathfrak{k}'$; that is, $\delta_O = \delta_A$.

To prove the simpler theorem formulated at the beginning of this proof we use only the fact that S is a continuous map of $\mathfrak{E}$ which carries distinct points into distinct points, and the fact that $\mathfrak{E}$ is a *convex set;* that is, the segment joining any two points of $\mathfrak{E}$ is contained in $\mathfrak{E}$. Again we use a fixed circle $\mathfrak{k}$, contained in $\mathfrak{E}$, with center O; let the radius be b. We have the parametric representation for $\mathfrak{k}$

$$x = b\cos 2\pi\lambda, \qquad y = b\sin 2\pi\lambda \qquad (0 \leq \lambda \leq 1); \tag{7.1}$$

we are assuming that O is the origin. Let O' have the (integral) order δ

relative to the image curve $\mathfrak{k}'$. Let the arbitrary closed curve $\mathfrak{C}$ in $\mathfrak{E}$, not through O, be given by

$$P = P(\lambda) \qquad \text{or} \qquad x = x(\lambda), \qquad y = y(\lambda) \qquad (0 \leq \lambda \leq 1),$$

and let $\phi = \phi(\lambda)$ be the azimuth of $\overrightarrow{OP}$, which varies continuously with λ (Fig. 6). We partition $\mathfrak{C}$ into subarcs,

$$0 \leq \lambda \leq \lambda_1, \qquad \lambda_1 \leq \lambda \leq \lambda_2, \qquad \ldots, \qquad \lambda_{r-1} \leq \lambda \leq 1,$$

so that any two values of ϕ, occurring on the same subarc $\mathfrak{C}_h : \lambda_{h-1} \leq \lambda \leq \lambda_h$,

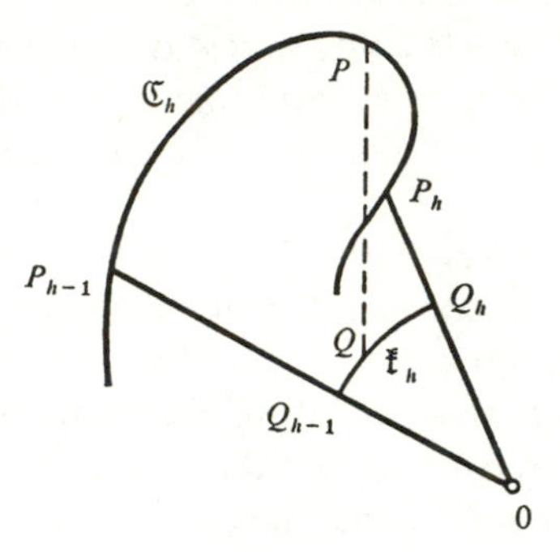

FIGURE 6

differ in absolute value by $< \frac{1}{4}$. The division point $P(\lambda_h)$ on the curve is denoted by P_h. With $\mathfrak{C}_h$ we associate the circular arc $\mathfrak{k}_h : Q_{h-1}\, Q_h$ of $\mathfrak{k}$ cut out by the rays OP_{h-1} and OP_h. We represent $\mathfrak{k}_h$ parametrically: $x = x_h(\lambda)$, $y = y_h(\lambda)$ so that when λ increases monotonically from λ_{h-1} to λ_h, the associated point (x, y) runs through that circular arc monotonically from Q_{h-1} to Q_h. The segment, in $\mathfrak{E}$, joining a point P on $\mathfrak{C}_h$ and a point Q on $\mathfrak{k}_h$ cannot pass through O; for if it did, the rays OP, OQ would form an angle of $180° = \frac{1}{2}$, which is impossible.

$$\mathfrak{K} = \mathfrak{k}_1 + \mathfrak{k}_2 + \cdots + \mathfrak{k}_r$$

is a closed curve which runs n times around $\mathfrak{k}$ in piecewise monotonic fashion. The image curve $\mathfrak{K}'$ of $\mathfrak{K}$ runs around $\mathfrak{k}'$ n times in a piecewise monotonic manner; hence O' has order $n\delta$ relative to $\mathfrak{K}'$.

If μ is a number, $0 \leq \mu \leq 1$, then the equations

$$\begin{aligned} x &= \mu x(\lambda) + (1 - \mu)\, x_h(\lambda) \\ y &= \mu y(\lambda) + (1 - \mu)\, y_h(\lambda) \end{aligned} \qquad (\lambda_{h-1} \leq \lambda \leq \lambda_h) \qquad h = 1, 2, \ldots, r,$$

represent a closed curve $\mathfrak{C}_\mu$, which does not pass through O, and whose image $\mathfrak{C}'_\mu$ does not pass through O'. If n'_μ is the order of O' relative to $\mathfrak{C}'_\mu$, then $\mathfrak{C}'_\mu$ and n'_μ vary continuously with μ. Since n'_μ is an integer, it must be a constant for $0 \leq \mu \leq 1$. For $\mu = 1$, $\mathfrak{C}_\mu = \mathfrak{C}$, and for $\mu = 0$, $\mathfrak{C}_\mu = \mathfrak{K}$. Thus we must have

$$\underset{\mathfrak{C}'}{\operatorname{ord}}(O') = \underset{\mathfrak{K}'}{\operatorname{ord}}(O') = n\delta .$$

A local topological map leaving O fixed in the xy-plane will be called *positive* if $\delta_O = +1$, *negative* if $\delta_O = -1$. The positive local topological maps form a subgroup Γ^+ of index 2 of the group Γ of all local topological maps. A surface of type Γ^+ is called *oriented.* According to our general considerations we have such a surface when with each point $\mathfrak{p}_0$ of the surface there is associated a class of local coordinate systems such that any system in the class may be obtained from *one* of them (x, y) by means of some transformation of the local group Γ^+. The result above, $\delta_A = \delta_O$, is important here; for the condition that $(x - x_1, y - y_1)$ is an admissible coordinate system for any point $P_1 = (x_1, y_1)$ sufficienctly close to $\mathfrak{p}_0$ must be satisfied. We can also describe the situation as follows. On an oriented surface there is associated with each point $\mathfrak{p}$ a unique positive sense of rotation $\vartheta(\mathfrak{p})$, which varies continuously with $\mathfrak{p}$; that is, under a local topological map into the plane, $\vartheta(\mathfrak{p})$ turns into one *fixed* sense of rotation in some neighborhood of the origin in the plane.

We still have to show that our earlier criterion, the sign of the functional determinant, that a local differentiable map be positive or negative agrees with our criterion for arbitrary topological maps. Taking the latter as our fundamental criterion, we remark first that for a differentiable map the sign δ depends only on the linear approximation to the map at the origin. If $a_1, b_1; a_2, b_2$ are the values of $\partial x^*/\partial x, \partial x^*/\partial y; \partial y^*/\partial x, \partial y^*/\partial y$ at the origin, then the linear approximation is

$$\xi_1 = a_1 x + b_1 y, \qquad \xi_2 = a_2 x + b_2 y . \tag{7.2}$$

The proof is as follows: δ is not only the order of O' relative to the image of the circle $\mathfrak{k}$, (7.1), but also relative to the image of the concentric circle $\mathfrak{k}_\varepsilon$ of radius εb; here ε is a positive factor which we shall let tend to zero. The ray from O, whose direction is given by the parameter λ in (7.1), meets the circle $\mathfrak{k}$ at a point P, the circle $\mathfrak{k}_\varepsilon$ at a point P_ε. Let P'_ε be the image of P_ε, let $\bar{P}$ be the image of P under the linear approximation (7.2), and let $\phi'_\varepsilon(\lambda)$, $\bar{\phi}(\lambda)$ be the angles for the directions OP'_ε and $O\bar{P}$. It is immediately seen that $\phi'_\varepsilon(\lambda)$ and $\bar{\phi}(\lambda)$ differ by an arbitrarily small amount, and in particular by less than $\frac{1}{4}$, if ε is sufficiently small. Thus if one follows the continuous change of $\phi'_\varepsilon(\lambda)$

and $\bar{\Phi}(\lambda)$, as λ increases from 0 to 1, it is seen that they both experience the same increase $\delta = \pm 1$.

Thus we need study only the affine map (7.2) and show that δ is the sign of the determinant $J = a_1 b_2 - a_2 b_1$. If we set

$$x = \cos 2\pi\phi, \qquad y = \sin 2\pi\phi, \tag{7.3}$$

then (x, y) runs through the unit circle; (ξ_1, ξ_2) traces an ellipse. We introduce polar coordinates r, ψ for (ξ_1, ξ_2): $\xi_1 = r \cos 2\pi\psi$, $\xi_2 = r \sin 2\pi\psi$, and follow the continuous change of ψ with ϕ. From the equations

$$r \cos 2\pi\psi = a_1 \cos 2\pi\phi + b_1 \sin 2\pi\phi,$$
$$r \sin 2\pi\psi = a_2 \cos 2\pi\phi + b_2 \sin 2\pi\phi$$

an elementary computation shows that

$$r^2\, d\psi = (a_1 b_2 - a_2 b_1)\, d\phi. \tag{7.4}$$

Since $1/r^2$ is a positive function of ϕ, then the increase δ experienced by ψ when ϕ increases from 0 to 1 is positive or negative as $a_1 b_2 - a_2 b_1$ is positive or negative; hence δ is the sign of this determinant. Thus we have proved that our two criteria, one applicable to arbitrary surfaces and the other to smooth surfaces, agree.

I find it pleasing (hübsch) to carry out all the details for such elementary considerations as the last. Therefore I derive in the simplest way an upper and a lower bound for the positive definite quadratic form

$$r^2 = Q(x, y) = \xi_1^2 + \xi_2^2 = (a_1 x + b_1 y)^2 + (a_2 x + b_2 y)^2.$$

The familiar inequality of Cauchy-Schwarz gives the upper bound

$$Q(x, y) \leq \{(a_1^2 + b_1^2) + (a_2^2 + b_2^2)\}(x^2 + y^2).$$

On the other hand, "completion of the square," familiar to the Arabs (or even the Babylonians?)

$$Q(x, y) = (a_1^2 + a_2^2)\left(x + \frac{a_1 b_1 + a_2 b_2}{a_1^2 + a_2^2}\, y\right)^2 + \frac{(a_1 b_2 - a_2 b_1)^2}{a_1^2 + a_2^2}\, y^2$$

leads to the inequality

$$(a_1^2 + a_2^2)\, Q \geq (a_1 b_2 - a_2 b_1)^2\, y^2.$$

From this and the inequality

$$(b_1^2 + b_2^2)\, Q \geq (a_1 b_2 - a_2 b_1)^2\, x^2,$$

which results from it by the interchange of (a, x) and (b, y), we get

$$(a_1^2 + a_2^2 + b_1^2 + b_2^2)\,Q(x, y) \geq (a_1 b_2 - a_2 b_1)^2 (x^2 + y^2).$$

Thus if we write for (7.4)

$$d\psi = \delta H(\phi)\,d\phi, \qquad \delta = \operatorname{sgn}(a_1 b_2 - a_2 b_1),$$

then $H(\phi)$ lies between the number

$$\frac{|a_1 b_2 - a_2 b_1|}{a_1^2 + a_2^2 + b_1^2 + b_2^2}$$

(lower bound) and its reciprocal. If the bounds of $H(\phi)$ did not enclose the number 1, then it would be impossible for

$$\int_0^1 H(\phi)\,d\phi$$

to have the same value, 1, as

$$\int_0^1 d\phi\,.$$

That Riemann surfaces are oriented surfaces in the topological sense may be seen as follows without using smooth surfaces. Under a conformal map

$$\tau = a_1 t + a_2 t^2 + \cdots \qquad (a_1 \neq 0) \tag{7.5}$$

the logarithmic differential of the local parameter behaves according to an equation

$$\frac{d\tau}{\tau} = dt\left\{\frac{1}{t} + c_0 + c_1 t + \ldots\right\}. \tag{7.6}$$

The integral

$$\frac{1}{2\pi i}\int \frac{dt}{t}$$

over a sufficiently small circle $\mathfrak{k}$ in the positive sense about the origin O of the complex t-plane has the value 1; the integral

$$\frac{1}{2\pi i}\int \frac{d\tau}{\tau}$$

over the same circle is the order δ of O relative to the image of $\mathfrak{k}$ under (7.5). But from (7.6), the value of this integral is also 1.

The concept of orientation is one of the simplest which one can follow through the stages: surface–smooth surface–Riemann surface. To be sure, it is true that Riemann surfaces are oriented by nature, and it would be possible to treat them without mentioning the theme of orientation. But we are interested in following function theory back to its topological roots.

§ 8. Covering surfaces

Let $\mathfrak{F}$, $\mathfrak{F}^*$ be two manifolds. By a continuous map $\mathfrak{p} \to \mathfrak{p}^*$ of $\mathfrak{F}$ into $\mathfrak{F}^*$ the first manifold is *imbedded* in the second. The image $\mathfrak{p}^*$ of $\mathfrak{p}$ is called the *place* of $\mathfrak{p}$ in $\mathfrak{F}^*$. Thus in the theory of surfaces in three-dimensional Euclidean space it is useful to regard the surface first as a (smooth) two-dimensional manifold consisting of elements *sui generis*, the "points of the surface," which do not have anything to do with space. Then each point $\mathfrak{p}$ of the surface is assigned a place in Euclidean space by means of a continuous map. The local coordinates on the surface, frequently denoted by u and v, are known as *Gaussian coordinates*; if x, y, z are Cartesian coordinates for the imbedding space, then "in the small" the imbedded surface has a parametric representation

$$x = x(u,v), \qquad y = y(u,v), \qquad z = z(u,v)$$

which assigns a surface point (u, v) to its place (x, y, z). Similarly, the curves on an arbitrary surface arise when one imbeds the one-dimensional manifold *kat' exochen* [from the outside], the interval $0 \leq \lambda \leq 1$, in the surface. Here we have one dimension less than in the previous example. It is by no means necessary that the dimension of the manifold in which one imbeds be higher than that of the imbedded one; if it is less, then one is apt to use the term *projection* instead of *imbedding*.

We are interested in the case where both dimension numbers have the same value, 2. Let $\mathfrak{F}$ and $\bar{\mathfrak{F}}$ be two given two-dimensional manifolds; for the moment we assume neither connectedness nor the axiom of countability. $\bar{\mathfrak{F}}$ becomes a *covering surface* over the base surface $\mathfrak{F}$ if to each point $\bar{\mathfrak{p}}$ of $\bar{\mathfrak{F}}$ there is associated a point $\mathfrak{p}$ of $\mathfrak{F}$, the trace of $\bar{\mathfrak{p}}$; we say also that $\bar{\mathfrak{p}}$ lies over $\mathfrak{p}$. We consider only *unbranched covering surfaces*. They are characterized by two properties: the imbedding map $S: \bar{\mathfrak{p}} \to \mathfrak{p}$ is continuous and neighborhood true; each point $\bar{\mathfrak{p}}_0$ of $\bar{\mathfrak{F}}$ has a neighborhood $\bar{\mathfrak{U}}$ such that distinct points of $\bar{\mathfrak{U}}$ never have the same trace points. If $\mathfrak{U}$ is the image of $\bar{\mathfrak{U}}$ under S, then $\mathfrak{U}$ is a domain consisting of interior points, and in particular contains $\mathfrak{p}_0$ in its interior; the map S^{-1} of $\mathfrak{U}$ onto $\bar{\mathfrak{U}}$ is continuous, and thus the pair S, S^{-1} is a topological map $\bar{\mathfrak{U}} \rightleftarrows \mathfrak{U}$.

As already betrayed by the terminology, our primary concern in not with the imbedded manifold $\bar{\mathfrak{F}}$, but rather with the base surface $\mathfrak{F}$ into which $\bar{\mathfrak{F}}$ is imbedded. The concept of *continuous continuation*, which appeared in the special form of analytic continuation in the initial §§ 1–3, finds its general topological formulation in the concept of unbranched covering surfaces. Namely, let γ be a curve $\mathfrak{p} = \mathfrak{p}(\lambda)$ on $\mathfrak{F}$, from $\mathfrak{o}$ to $\mathfrak{a}$, and $\bar{\mathfrak{o}}$ a point of $\bar{\mathfrak{F}}$ which lies over $\mathfrak{o}$. We wish to follow the continuous motion of a point $\bar{\mathfrak{p}}$ on $\bar{\mathfrak{F}}$, $\bar{\mathfrak{p}} = \bar{\mathfrak{p}}(\lambda)$, which starts at $\bar{\mathfrak{o}}$ and whose trace follows the given path $\mathfrak{p} = \mathfrak{p}(\lambda)$. I claim: *either* this is possible along all of γ, and the continuous curve $\bar{\gamma}$ on $\bar{\mathfrak{F}}$, which starts at $\bar{\mathfrak{o}}$ and has trace γ, is uniquely determined; *or* there exists a positive threshold $\Lambda_0 \leq 1$, such that this continuation is possible over the part of the curve γ determined by $0 \leq \lambda < \Lambda_0$, but continuation to the point $\lambda = \Lambda_0$ is impossible. One can grasp the last case as follows. If one follows from $\bar{\mathfrak{o}}$ the continuous change of a point $\bar{\mathfrak{p}}$ on $\bar{\mathfrak{F}}$, whose trace on $\mathfrak{F}$ follows the curve γ, then before reaching the end one bumps into a *border* of the covering surface over the point $\mathfrak{p}(\Lambda_0)$ of $\mathfrak{F}$. If only the first case arises, whatever the point $\bar{\mathfrak{o}}$ on $\bar{\mathfrak{F}}$ and whatever the curve γ on $\mathfrak{F}$, starting at the trace $\mathfrak{o}$ of $\bar{\mathfrak{o}}$, may be, then we will call the unbranched covering surface $\bar{\mathfrak{F}}$ *unlimited.*

We prove these statements first in the small. Let $\bar{\mathfrak{o}}$ be a point of $\bar{\mathfrak{F}}$ with trace $\mathfrak{o}$. There exists a neighborhood $\bar{\mathfrak{U}}$ of $\bar{\mathfrak{o}}$ such that if two points $\bar{\mathfrak{p}}_1$ and $\bar{\mathfrak{p}}_2$ in $\bar{\mathfrak{U}}$ have the same trace $\mathfrak{p}$, then they are the same. This neighborhood is mapped topologically onto a domain $\mathfrak{U}$ of $\mathfrak{F}$ by $S: \bar{\mathfrak{p}} \to \mathfrak{p}$. If $\mathfrak{u}$ is a neighborhood of $\mathfrak{o}$, contained in $\mathfrak{U}$, then we have a unique continuous map S^{-1}: $\mathfrak{p} \to \bar{\mathfrak{p}}$ of $\mathfrak{u}$ into $\bar{\mathfrak{F}}$ under which $\mathfrak{o}$ goes into $\bar{\mathfrak{o}}$, and each point $\mathfrak{p}$ goes into a point $\bar{\mathfrak{p}}$ over $\mathfrak{p}$. If $\bar{\mathfrak{o}}_1$ and $\bar{\mathfrak{o}}_2$ are two distinct points over $\mathfrak{o}$, then they possess disjoint neighborhoods $\bar{\mathfrak{U}}_1$ and $\bar{\mathfrak{U}}_2$; thus a point $\bar{\mathfrak{p}}_1$ in $\bar{\mathfrak{U}}_1$ and a point $\bar{\mathfrak{p}}_2$ in $\bar{\mathfrak{U}}_2$ are never the same, whether they possess the same trace or not.

From what has been said it follows that: (a) If one has succeeded in constructing the curve $\bar{\gamma}$, starting at $\bar{\mathfrak{o}}$, with trace γ, for the subinterval $0 \leq \lambda \leq \lambda_1$ (where $0 < \lambda_1 < 1$), then it is possible to continue $\bar{\gamma}$ beyond λ_1. For by the above we can choose a positive ε such that we get a curve $\bar{\mathfrak{p}}_1(\lambda)$ over $\mathfrak{p}(\lambda)$ for $\lambda_1 - \varepsilon \leq \lambda \leq \lambda_1 + \varepsilon$ for which $\bar{\mathfrak{p}}_1(\lambda_1) = \bar{\mathfrak{p}}(\lambda_1)$. But this $\bar{\mathfrak{p}}_1(\lambda)$ coincides with $\bar{\mathfrak{p}}(\lambda)$ for $\lambda_1 - \varepsilon \leq \lambda \leq \lambda_1$; thus in fact we have a continuation to the value $\lambda = \lambda_1 + \varepsilon$.

(b) If $\bar{\mathfrak{p}}_1(\lambda)$ and $\bar{\mathfrak{p}}_2(\lambda)$ are two curves over $\mathfrak{p}(\lambda)$ for $0 \leq \lambda \leq 1$, then either $\bar{\mathfrak{p}}_1(\lambda) = \bar{\mathfrak{p}}_2(\lambda)$ for *each* value λ in the unit interval, or $\bar{\mathfrak{p}}_1(\lambda) \neq \bar{\mathfrak{p}}_2(\lambda)$. For about each value λ_0 one can find an interval such that the relation $\bar{\mathfrak{p}}_1 = \bar{\mathfrak{p}}_2$ or $\bar{\mathfrak{p}}_1 \neq \bar{\mathfrak{p}}_2$ for λ_0 entails the same equality or inequality for *all* values λ in the interval. Using the Heine-Borel principle, one can partition the curve into a finite number of arcs so that on each arc either $\bar{\mathfrak{p}}_1(\lambda) = \bar{\mathfrak{p}}_2(\lambda)$ everywhere on the arc or $\bar{\mathfrak{p}}_1(\lambda) \neq \bar{\mathfrak{p}}_2(\lambda)$ everywhere on the arc. The claim (b) follows from piecing these arcs together at the division points. The fact that

$\bar{\mathfrak{p}}_1(0) = \bar{\mathfrak{p}}_2(0) = \bar{\mathfrak{o}}$ rules out the second possibility, $\bar{\mathfrak{p}}_1(\lambda) \neq \bar{\mathfrak{p}}_2(\lambda)$ for $0 \leq \lambda \leq 1$.

(c) If one has a curve $\bar{\mathfrak{p}}_1(\lambda)$, starting at $\bar{\mathfrak{o}}$, and over $\mathfrak{p}(\lambda)$, for $0 \leq \lambda \leq \lambda_1$, and a second such curve $\bar{\mathfrak{p}}_2(\lambda)$ for $0 \leq \lambda \leq \lambda_2$, and if $\lambda_1 \leq \lambda_2$, then $\bar{\mathfrak{p}}_1(\lambda)$ and $\bar{\mathfrak{p}}_2(\lambda)$ coincide in the interval $0 \leq \lambda \leq \lambda_1$. Thus $\bar{\mathfrak{p}}_2(\lambda)$ is a continuation of $\bar{\mathfrak{p}}_1(\lambda)$ from λ_1 to λ_2.

(d) From these three facts the claimed alternative follows: the continuation along γ, starting at $\bar{\mathfrak{o}}$, either succeeds along all of γ, or along an open piece $0 \leq \lambda < \Lambda_0$ (with $0 < \Lambda_0 \leq 1$) of γ; in both cases the continuation is unique.

For a connected, unbranched, and unlimited covering surface $\bar{\mathfrak{F}}$ over a connected two-dimensional manifold $\mathfrak{F}$, we shall use the shorter name, *perfect covering surface*. It is *one-sheeted* if there is just one point of $\bar{\mathfrak{F}}$ over each point of $\mathfrak{F}$; then $\bar{\mathfrak{F}}$ is not essentially different from $\mathfrak{F}$. $\mathfrak{F}$ is called *simply connected* if it has no perfect covering surfaces except the one-sheeted ones.[18] If $\bar{\mathfrak{F}}$ is an unbranched unlimited covering surface over $\mathfrak{F}$ and if $\mathfrak{G}$ is a simply connected domain on $\mathfrak{F}$, then the set of points on $\bar{\mathfrak{F}}$ which lie over points of $\mathfrak{G}$ falls apart into single *sheets* which are not connected with each other. Each sheet is mapped topologically onto $\mathfrak{G}$ by the trace map $\bar{\mathfrak{p}} \to \mathfrak{p}$. A sheet is uniquely determined as soon as the point $\bar{\mathfrak{a}}$ in the sheet over *one* point $\mathfrak{a}$ of $\mathfrak{G}$ is given.

Examples of perfect covering surfaces. (1) If one wraps a cylindrical sausage of length h in an infinitely long strip of paper of width h, then this paper strip becomes an infinitely many sheeted perfect covering surface over the cylindrical jacket of the sausage. (2) By associating with a point P of the unit sphere in three-dimensional space the line through the origin O and P, one turns the sphere into a two-sheeted perfect covering surface over the projective plane (see Examples 3 and 4 in § 5). (3) Similarly, the parallel strip in Examples 5 of § 5 becomes a covering surface of the Möbius strip if one identifies the points of the parallel strip which are equivalent under the discrete group of paddle motions (5.2) described there. (4) There is an analogous relation between the torus of Example 6 and the ϕ, ψ-plane; here two points (ϕ_1, ψ_1) and (ϕ_2, ψ_2) of the plane have the same trace on the torus if $\phi_1 \equiv \phi_2 \pmod{a}$ and $\psi_1 \equiv \psi_2 \pmod{b}$.

The interior $\mathfrak{E}$ *of the unit circle in the xy-plane is a simply connected surface.* We conduct the proof in a way that will be familiar to most who have read proofs of the Cauchy integral theorem. Let then $\bar{\mathfrak{E}}$ be a perfect covering surface over $\mathfrak{E}$ and let $\bar{O}$ be a point of $\bar{\mathfrak{E}}$ over the center O of $\mathfrak{E}$. For each

---

[18]) This definition singles out that property of simply connected surfaces which is decisive in function theoretic applications.

segment $\gamma = OP$ in $\mathfrak{E}$ we can find the curve in $\bar{\mathfrak{E}}$, starting at $\bar{O}$, whose trace is γ. It ends at a *uniquely* determined point $\bar{P}$ of $\bar{\mathfrak{E}}$ over P. The inverse of this single-valued mapping $T: P \to \bar{P}$ is the imbedding $S: \bar{P} \to P$. It remains to be shown that T is continuous; that is, for an arbitrary point A of $\mathfrak{E}$ one can give a circular neighborhood K of A such that for each point P in this neighborhood, the curve, starting at $\bar{A}$, over the segment AP ends in the same point $\bar{P}$ at which one arrives by starting at $\bar{O}$ and following the curve over the segment OP. We know that each point Q of the segment OA has a circular neighborhood $\mathfrak{u}(Q)$ such that the points of $\mathfrak{u}(Q)$ are carried into points above them in $\bar{\mathfrak{E}}$ by a continuous map which carries Q into $\bar{Q}$. The segment OA may be partitioned into finitely many partial segments

$$\sigma_i = A_{i-1}A_i \qquad (i = 1, \ldots, n;\ A_0 = O,\ A_n = A),$$

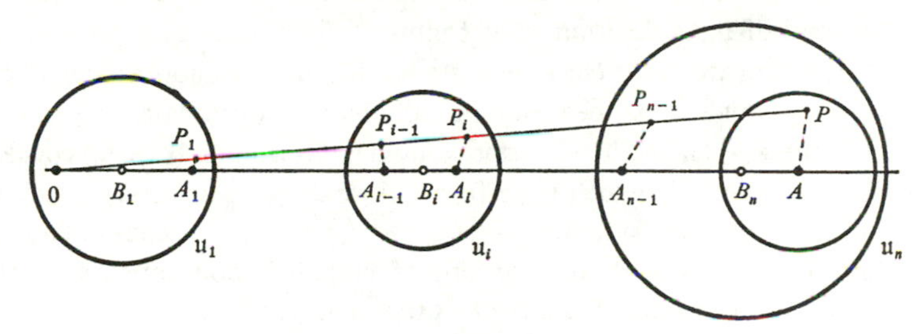

Fig. 7. Simple connectivity of the disc.

with a point B_i specified on each σ_i, such that σ_i is contained in the neighborhood $\mathfrak{u}_i = \mathfrak{u}(B_i)$ associated with B_i. If P is sufficiently close to A, that is, in a certain circle concentric with $\mathfrak{u}(A)$, then the segment OP may be partitioned by splitting points P_i (with $P_0 = O$, $P_n = P$) so that always P_{i-1} and P_i lie in $\mathfrak{u}_i$ (just as do A_{i-1} and A_i) ($i = 1, \ldots, n$). It follows, if we trace the closed quadrilateral $B_iA_iP_iP_{i-1}A_{i-1}(B_i)$, starting at B_i, that the quadrilateral lying over it and starting at $\bar{B}_i$ is also closed. Summation over $i = 1, \ldots, n$ justifies our claim that continuation along the closed path $OAPO$ on $\bar{\mathfrak{E}}$ leads back to the starting point $\bar{O}$.

Obviously the same argument serves for any *convex* domain $\mathfrak{G}$ in the xy-plane. Thus the interior of a square or triangle is also simply connected. All neighborhoods on a two-dimensional manifold are simply connected, since they are topological images of the disc $\mathfrak{E}$. This fact simplifies the investigation of perfect covering surfaces significantly: one knows now that for each neighborhood $\mathfrak{U}$ of a point $\mathfrak{p}_0$ on $\mathfrak{F}$ and a point $\bar{\mathfrak{p}}_0$ over $\mathfrak{p}_0$ on $\bar{\mathfrak{F}}$

there is a uniquely determined domain $\bar{\mathfrak{U}}$ on $\bar{\mathfrak{F}}$, containing $\bar{\mathfrak{p}}_0$, which is mapped topologically on $\mathfrak{U}$ by the imbedding relation.

We have seen that with each point $\mathfrak{p}$ of a two-dimensional manifold $\mathfrak{F}$ one can associate one of two opposite senses of rotation $\pm\vartheta$. The elements $\bar{\mathfrak{p}}$ = "point $\mathfrak{p}$ of $\mathfrak{F}$ + sense of rotation at $\mathfrak{p}$" can be regarded as the points of a two-sheeted, unbranched, unlimited covering surface $\bar{\mathfrak{F}}$ over $\mathfrak{F}$ in which the element $\bar{\mathfrak{p}}$ is assigned the point $\mathfrak{p}$ of $\mathfrak{F}$ as its trace point. How the concept of neighborhood is carried over to $\bar{\mathfrak{F}}$ is naturally essential. For this purpose we repeat: a point $\mathfrak{p}_0$ of $\mathfrak{F}$ together with a coordinate system (x, y) at $\mathfrak{p}_0$ defines a point $\bar{\mathfrak{p}}_0$ of $\bar{\mathfrak{F}}$ over $\mathfrak{p}_0$; $\mathfrak{p}_0'$ together with the local coordinate system (x', y') at $\mathfrak{p}_0'$ is the same point if and only if $\mathfrak{p}_0' = \mathfrak{p}_0$ and (x', y') arise from (x, y) by a *positive* topological transformation. Accordingly, of all the local coordinate systems (x, y) at $\mathfrak{p}_0$, the "counterclockwise" ones belong to one of the two points $\bar{\mathfrak{p}}_0$ over $\mathfrak{p}_0$, the "clockwise" ones belong to the other. A neighborhood of $\bar{\mathfrak{p}}_0$ is determined as follows. We choose local coordinates (x, y) at $\mathfrak{p}_0$ which are in the class belonging to $\bar{\mathfrak{p}}_0$, and we choose a neighborhood $\mathfrak{U}$ of $\mathfrak{p}_0$ which is mapped topologically by the coordinates x, y onto a domain in the xy-plane. The corresponding neighborhood $\bar{\mathfrak{U}}$ of $\bar{\mathfrak{p}}_0$ consists then of those points $\bar{\mathfrak{p}}_1$ which arise from associating an arbitrary point $\mathfrak{p}_1$ of $\mathfrak{U}$ with the coordinate system $(x - x_1, y - y_1)$, (x_1, y_1 = values of x, y at $\mathfrak{p}_1$). One must verify that this concept of neighborhood satisfies all the conditions. But this follows from the fact that a local topological coordinate transformation $(x, y) \to (x^*, y^*)$ which is positive or negative at the origin has the same character at every point of some neighborhood of the origin; recall the equation $\delta_A = \delta_O$ on page 45! Therefore for two neighborhoods of $\bar{\mathfrak{p}}_0$ there exists one which is contained in both; and the two points $\bar{\mathfrak{p}}_0 = \mathfrak{p}_0^+$ and $\mathfrak{p}_0^-$ over $\mathfrak{p}_0$ possess a pair of disjoint neighborhoods.

If the covering surface $\bar{\mathfrak{F}}$, of the connected surface $\mathfrak{F}$, which distinguishes between the two senses of rotation at points $\mathfrak{p}_0$ of $\mathfrak{F}$ is not connected, but consists of two sheets, then the surface $\mathfrak{F}$ may be turned into an oriented surface in two ways; in this case the surface is said to be *orientable* or *two sided.* But if $\bar{\mathfrak{F}}$ is connected, then $\mathfrak{F}$ is not orientable in any way: no coherent sense of rotation can be specified on $\mathfrak{F}$; that is, it is impossible to associate a sense of rotation with each point $\mathfrak{p}$ of $\mathfrak{F}$ so that it depends continuously on $\mathfrak{p}$. Naturally, the covering surface $\bar{\mathfrak{F}}$ is, as an oriented surface, two sided. Furthermore, it follows from the definition that *a simply connected surface is necessarily two sided.*

Examples of one-sided surfaces: the projective plane and the Möbius strip. The projective plane arises from the sphere by identifying each pair of antipodal points. Above (p. 37) we described a definite orientation of the unit sphere. Under the map which carries each point into its antipodal point this orientation is turned into the other orientation. Thus the projective plane is one sided, and the sphere can be regarded as the covering surface which makes out of one point two points with opposite senses of rotation attached. Similarly, the Möbius strip is one sided. In the plane strip, which appeared in § 5 as a covering surface of the Möbius strip, draw the midline. It corresponds to a closed curve on the Möbius strip, and a fly which crawls along this curve on the strip has arrived on the other side of the sheet when it first returns to its starting point; and it hasn't crawled over the edge. Here we talk of the "*side*" in the same sense as when we say that we bescrawl first one side and then the other of a sheet of paper. The plane has two separate sides, while the two sides of the Möbius strip are not separated. The two-sheeted covering surface $\bar{\mathfrak{F}}$ represents, as it were, the two *sides* of the surface $\mathfrak{F}$.[19]

Let $\mathfrak{F}$ be an oriented surface, $\mathfrak{p}_0$ a point on $\mathfrak{F}$, and $\mathfrak{U}$ an arbitrary neighborhood of $\mathfrak{p}_0$. Let $\mathfrak{U}$ be mapped topologically onto $\mathfrak{E}$, the interior of the unit circle in the xy-plane, so that $\mathfrak{p}_0$ goes into the origin and so that (x, y) establishes the positive sense of rotation at $\mathfrak{p}_0$. From the fact that $\mathfrak{E}$, and hence $\mathfrak{U}$, is simply connected, it follows that at each point $\mathfrak{p}_1$ of $\mathfrak{U}$ the local coordinate system $(x - x_1, y - y_1)$ establishes the right positive sense of rotation.

We assume now that the base surface $\mathfrak{F}$ is not only a *connected* two-dimensional manifold but that it also satisfies the *axiom of countability*. We shall prove that *every perfect covering surface* $\bar{\mathfrak{F}}$ *over* $\mathfrak{F}$ *satisfies this same axiom*. Our nomenclature, which from the beginning has been covering *surface* instead of covering *manifold*, has, so to speak, taken this fact for granted.

Let $\bar{\mathfrak{o}}$ be a fixed point of $\bar{\mathfrak{F}}$ over $\mathfrak{o}$. Every point $\bar{\mathfrak{p}}$ on $\bar{\mathfrak{F}}$ is uniquely determined by (1) its trace $\mathfrak{p}$ on $\mathfrak{F}$, and (2) a curve γ joining $\mathfrak{o}$ to $\mathfrak{p}$. Namely, $\bar{\mathfrak{p}}$ is obtained as the endpoint of the curve $\bar{\gamma}$, starting at $\bar{\mathfrak{o}}$, with trace γ. It is to be expected that the continuation along two such curves γ leads to the same endpoint $\bar{\mathfrak{p}}$ in case the two paths run close enough together. From this it should follow that only countably many different points of the covering

---

[19]) On a surface located in space we distinguish the two sides of the surface at a point by the two directions which may be assigned to the normal to the surface at this point. A sense of rotation on the surface may be joined with the direction of the normal so that together they constitute a "left-handed screw." If we replace the direction of the normal by this sense of rotation on the surface, we become independent of the imbedding of the surface in space. Felix Klein accomplished this step in the definition of one sided and two sided: *Math. Ann.*, **9** (1876) p. 479.

surface can lie over $\mathfrak{p}$. The result is analogous to the proof in § 3 showing that analytic continuation of a given function element can lead to only countably many function elements with a given center $z = a$. The greater generality of the concepts we are now using forces certain modifications in the process. At any rate, it is enough to show that by continuation along *closed paths* γ on $\mathfrak{F}$, which run from $\mathfrak{o}$ to $\mathfrak{o}$, one obtains on the surface $\tilde{\mathfrak{F}}$, by starting at $\tilde{\mathfrak{o}}$, only countably many different points over $\mathfrak{o}$.

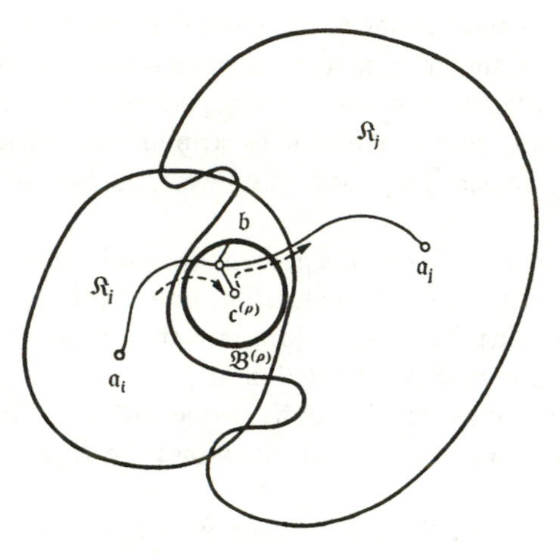

Fig. 8. Illustration of the proof that a perfect covering surface satisfies the axiom of countability.

For each point $\mathfrak{p}$ we determine a neighborhood $\mathfrak{U}(\mathfrak{p})$ and map it topologically onto the unit disc $\mathfrak{E}$ of the xy-plane. We shrink that disc to a closed disc $x^2 + y^2 \leq \rho^2$ of radius $\rho < 1$; we denote by $K(\mathfrak{p})$ (or $\mathfrak{K}(\mathfrak{p})$) the preimage of that closed disc (or its interior). Since $\mathfrak{F}$ satisfies the axiom of countability, we can choose a countable set of points $\mathfrak{a}_1, \mathfrak{a}_2, \ldots$ such that the associated $\mathfrak{K}_i = \mathfrak{K}(\mathfrak{a}_i)$ cover the whole surface. Set $\mathfrak{U}(\mathfrak{a}_i) = \mathfrak{U}_i$, $K(\mathfrak{a}_i) = K_i$. For $\mathfrak{a}_1$ we take the point $\mathfrak{o}$. Now let γ, $\mathfrak{p} = \mathfrak{p}(\lambda)$ $(0 \leq \lambda \leq 1)$, be a given closed curve on $\mathfrak{F}$ which runs from $\mathfrak{o}$ to $\mathfrak{o}$. With each value λ of the unit interval we associate a neighborhood of $\mathfrak{p}(\lambda)$ which is contained in one of the $\mathfrak{K}_i$; it follows that γ may be split into finitely many subarcs $\gamma', \gamma'', \ldots$, each contained in some $\mathfrak{K}_i$. These arcs are cyclically ordered; after the last comes the first again. Two successive arcs which lie in the same $\mathfrak{K}_i$ we make into one arc. We replace the arc $\mathfrak{b}'\mathfrak{b}$ in $\mathfrak{K}_i$ by a path $\mathfrak{b}'\mathfrak{a}_i\mathfrak{b}$ in $\mathfrak{K}_i$ through the center of $\mathfrak{K}_i$ (for example, the path which in the image consists of the two linear segments

$\mathfrak{b}'\mathfrak{a}_i, \mathfrak{a}_i\mathfrak{b}$). Continuation on $\mathfrak{F}$ along one or the other path leads to the same result since $\mathfrak{U}_i$ is simply connected. After this alteration, γ consists of arcs $\mathfrak{a}_i\mathfrak{b}\mathfrak{a}_j$ whose first part $\mathfrak{a}_i\mathfrak{b}$ lies in $\mathfrak{K}_i$ and whose second $\mathfrak{b}\mathfrak{a}_j$ lies in $\mathfrak{K}_j$. The point $\mathfrak{b}$ lies in the intersection $\mathfrak{K}_i \cap \mathfrak{K}_j$. Each point of the compact set $K_i \cap K_j$ has a neighborhood $\mathfrak{B}$ which is contained in both $\mathfrak{U}_i$ and $\mathfrak{U}_j$. From these one can choose finitely many $\mathfrak{B}^{(\rho)}$ ($\rho = 1, \ldots, r$) which cover $K_i \cap K_j$. One of these "bridges" $\mathfrak{B}^{(\rho)}$ contains $\mathfrak{b}$. Join $\mathfrak{b}$ inside $\mathfrak{B}^{(\rho)}$ with the center $\mathfrak{c}^{(\rho)}$ of this bridge and sandwich this path $\mathfrak{b}\mathfrak{c}^{(\rho)}$, traced back and forth, between $\mathfrak{a}_i\mathfrak{b}$ and $\mathfrak{b}\mathfrak{a}_j$. This has no effect on the continuation, but the path $\mathfrak{a}_i\mathfrak{b}\mathfrak{a}_j$ has now changed to $\mathfrak{a}_i\mathfrak{c}^{(\rho)}\mathfrak{a}_j$, where the first part $\mathfrak{a}_i\mathfrak{c}^{(\rho)}$ joins, inside $\mathfrak{U}_i$, the center of $\mathfrak{K}_i$ with the center of $\mathfrak{B}^{(\rho)}$, and the second part $\mathfrak{c}^{(\rho)}\mathfrak{a}_j$ runs in $\mathfrak{U}_j$ from $\mathfrak{c}^{(\rho)}$ to the center of $\mathfrak{K}_j$. For continuation on $\mathfrak{F}$, the detailed structure of these individual paths does not matter, for $\mathfrak{U}_i$ and $\mathfrak{U}_j$ are simply connected. For instance, one can choose them such that they become linear segments in the topological images of $\mathfrak{U}_i$ and $\mathfrak{U}_j$ in the plane. The point at which one arrives by continuation along γ, starting at $\bar{\mathfrak{o}}$, is thus uniquely determined by a finite cyclical chain, $\mathfrak{K}_i, \mathfrak{K}_j \ldots$, which begins with $\mathfrak{K}_1$ and one of the $r = r_{ij}$ available bridges in each adjacent pair $\mathfrak{K}_i, \mathfrak{K}_j$. Assigning the number $i = i_1 + \cdots + i_n$ as the "height" of such a "chain $\mathfrak{K}_{i_1}, \ldots, \mathfrak{K}_{i_n}$ with bridges," one sees that there are only a finite number of such entities with a given height i. Thus they can be counted, and one finds that only a countable number of points of $\mathfrak{F}$ can lie over $\mathfrak{o}$.

A perfect covering surface over a perfect covering surface of the base surface $\mathfrak{F}$ is a perfect covering surface over $\mathfrak{F}$.

A perfect covering surface $\bar{\mathfrak{F}}$ over $\mathfrak{F}$ is called *normal* if it never happens that, of two curves on $\bar{\mathfrak{F}}$ which possess the same closed trace curve on $\mathfrak{F}$, one is closed and the other not closed. A topological map S of $\bar{\mathfrak{F}}$ onto itself that carries each point $\bar{\mathfrak{p}}$ of $\bar{\mathfrak{F}}$ into a point $\bar{\mathfrak{p}}S$ over it (that is, $\bar{\mathfrak{p}}$ and $\bar{\mathfrak{p}}S$ have the same trace $\mathfrak{p}$ in $\mathfrak{F}$) is called a *cover transformation* of $\bar{\mathfrak{F}}$. The cover transformations obviously form a group. The relation of the two concepts just defined appears in the following theorem.

Let $\bar{\mathfrak{o}}, \bar{\mathfrak{o}}'$ be two points over each other on the normal covering surface $\bar{\mathfrak{F}}$. Then there exists exactly one cover transformation S of $\bar{\mathfrak{F}}$ which carries $\bar{\mathfrak{o}}$ into $\bar{\mathfrak{o}}'$.

Let $\bar{\mathfrak{p}}$ be a point of $\bar{\mathfrak{F}}$; join $\bar{\mathfrak{p}}$ with $\bar{\mathfrak{o}}$ by a curve $\bar{\gamma}$. Then it follows by continuation that the desired cover transformation (if it exists) must carry $\bar{\mathfrak{p}}$ into the endpoint $\bar{\mathfrak{p}}'$ of the curve $\bar{\gamma}'$ which starts at $\bar{\mathfrak{o}}'$ and has the same trace curve γ as $\bar{\gamma}$. This proves the uniqueness of S. If $\bar{\gamma}_1$ is a second curve from $\bar{\mathfrak{o}}$ to $\bar{\mathfrak{p}}$, then the curve $\bar{\gamma} - \bar{\gamma}_1$, which consists of $\bar{\gamma}$ followed by $\bar{\gamma}_1$ with its direction reversed, is closed. *In case the covering surface* $\mathfrak{F}$ *is normal,* then also the curve $(\bar{\gamma} - \bar{\gamma}_1)'$ which starts at $\bar{\mathfrak{o}}'$ and has the same trace as $\bar{\gamma} - \bar{\gamma}_1$, is closed. That is $\bar{\gamma}_1'$ leads from $\bar{\mathfrak{o}}'$ to the same point $\bar{\mathfrak{p}}'$ that $\bar{\gamma}'$ does. Thus the image point $\bar{\mathfrak{p}}'$ of $\bar{\mathfrak{p}}$ is uniquely determined, and the desired cover transformation $S: \bar{\mathfrak{p}} \to \bar{\mathfrak{p}}'$ has been constructed.

Let $\bar{\alpha}$ be a curve from $\bar{\mathfrak{o}}$ to $\bar{\mathfrak{o}}'$; its trace α is closed, and one can call $\bar{\mathfrak{o}}'$ the point which one obtains by following α on $\bar{\mathfrak{F}}$, starting at $\bar{\mathfrak{o}}$. If $\bar{\mathfrak{o}}$ is fixed on $\bar{\mathfrak{F}}$ once and for all, then S is determined by α and may be denoted by S_α. If now $\bar{\mathfrak{p}}$ also lies over $\mathfrak{o}$, then γ is a curve from $\mathfrak{o}$ to $\mathfrak{o}$ and $\bar{\mathfrak{o}}$ goes into $\bar{\mathfrak{p}}$ under the cover transformation $T = S_\gamma$. The point $\bar{\mathfrak{p}}' = \bar{\mathfrak{p}}S = \bar{\mathfrak{o}}TS$ is obtained by following the path $\alpha + \gamma$ on $\bar{\mathfrak{F}}$, starting at $\bar{\mathfrak{o}}$, which leads first to $\bar{\mathfrak{o}}'$, and then to $\bar{\mathfrak{p}}'$:

$$S_\gamma S_\alpha = S_{\alpha+\gamma}. \tag{8.1}$$

Thus addition of the paths starting and ending at $\mathfrak{o}$ corresponds to the composition of the associated cover transformations *in the reverse order.*

The group of cover transformations, regarded as an abstract group, expresses purely and completely everything in the relation between the normal covering surface $\bar{\mathfrak{F}}$ and the base surface $\mathfrak{F}$ which has the character of analysis situs. This group is also called the *Galois group* of $\bar{\mathfrak{F}}$. It is in fact the analog of the Galois group of a normal algebraic field (of finite degree) over a base field.

Among all the perfect covering surfaces which belong to a given surface $\mathfrak{F}$ there is one which is the "strongest."[20] It is characterized by the statement: a curve $\tilde{\gamma}$ on it is closed only if all curves on all perfect covering surfaces of $\mathfrak{F}$ which have the same trace as $\tilde{\gamma}$ are closed. As a consequence of this property, the "*universal covering surface*" $\tilde{\mathfrak{F}}$ is *normal*; the group of its cover transformations, regarded as an abstract group, the so-called fundamental group of $\mathfrak{F}$, is an analysis situs invariant of the base surface $\mathfrak{F}$. Also $\tilde{\mathfrak{F}}$ is simply connected. For if $\tilde{\mathfrak{F}}^*$ were a more than one-sheeted perfect covering surface over $\tilde{\mathfrak{F}}$, then one could draw an unclosed curve on $\tilde{\mathfrak{F}}^*$ whose trace on $\tilde{\mathfrak{F}}$ would be closed. Since $\tilde{\mathfrak{F}}^*$ is also a perfect covering surface over $\mathfrak{F}$, this would immediately contradict the characteristic property of $\tilde{\mathfrak{F}}$.

The universal covering surface may be defined as follows. If $\mathfrak{p}_0$ is a fixed point of $\mathfrak{F}$ then every curve γ starting at $\mathfrak{p}_0$ defines a "point of $\tilde{\mathfrak{F}}$" which we say lies over the endpoint of γ. Two such curves γ, γ' define the *same* point of $\tilde{\mathfrak{F}}$ if and only if on every perfect covering surface over $\mathfrak{F}$ every pair of curves which start at the same point and have traces γ, γ' end at the same point. Let γ_0 be a curve on $\mathfrak{F}$ from $\mathfrak{p}_0$ to $\mathfrak{p}$ which defines the point $\tilde{\mathfrak{p}}$ on $\tilde{\mathfrak{F}}$,

---

[20]) Poincaré, *Bulletin de la société mathématique de France,* **11** (1883) 113–114.

and let $\mathfrak{U}$ be a neighborhood of $\mathfrak{p}$ on $\mathfrak{F}$. If I attach to γ_0 all possible curves γ in $\mathfrak{U}$ which start at $\mathfrak{p}$, then I say that the points of $\tilde{\mathfrak{F}}$ defined by all these curves $\gamma_0 + \gamma$ form a "neighborhood" $\tilde{\mathfrak{U}}$ of $\tilde{\mathfrak{p}}$. Since $\mathfrak{U}$ is simply connected, there is just one point of $\tilde{\mathfrak{U}}$ over each point of $\mathfrak{U}$; hence our concept of "neighborhood" satisfies the conditions stated in § 4.

The significance of covering surfaces for complex function theory results from the fact that every perfect covering surface of a *Riemann surface* $\mathfrak{F}$ is itself a Riemann surface without more ado. For let $t(\mathfrak{p})$ be a local parameter at a point $\mathfrak{p}_0$ of $\mathfrak{F}$ and assign to each point of the covering surface over $\mathfrak{p}$ the same function value $t(\mathfrak{p})$; thus one obtains a local parameter at each point of $\tilde{\mathfrak{F}}$ over $\mathfrak{p}_0$. Similarly from every function on $\mathfrak{F}$, free of essential singularities, one obtains a function of the same character on $\tilde{\mathfrak{F}}$. Every function on an *arbitrary* perfect covering surface $\tilde{\mathfrak{F}}$ reappears as a uniform function on the *universal* covering surface $\tilde{\mathfrak{F}}$; so a study of the universal covering surface is capable of replacing, to some extent, the study of all other not so strong covering surfaces. Among the functions on $\tilde{\mathfrak{F}}$, those which are functions f on the base surface $\mathfrak{F}$ are characterized by the property that they remain invariant under the group of cover transformations of $\tilde{\mathfrak{F}}$; that is, they satisfy the identity

$$f(\tilde{\mathfrak{p}}) = f(\tilde{\mathfrak{p}}T),$$

where $T: \tilde{\mathfrak{p}} \to \tilde{\mathfrak{p}}T$ is any of these cover transformations.

The function theoretic exploitation of the simple connectivity of a surface rests on the following *monodromy theorem*.

If $\mathfrak{F}$ is a simply connected surface, if

$$z = a_{-m}t^{-m} + \cdots + a_0 + a_1 t + a_2 t^2 + \cdots \qquad (t = \text{local parameter at } \mathfrak{p}_0)$$

is a function element on $\mathfrak{F}$ at $\mathfrak{p}_0$, and if in the analytic continuation of z along arbitrary paths on $\mathfrak{F}$ one never meets critical points other than ordinary poles, then these continuations form a uniform (single-valued) function on $\mathfrak{F}$, regular analytic except for poles.

Proof. One can define a perfect covering surface over $\mathfrak{F}$ in a manner analogous to the definition of the universal covering surface. A curve on this covering surface, whose trace on $\mathfrak{F}$ runs from $\mathfrak{p}_0$ to $\mathfrak{p}_0$, shall be closed if and only if the continuation of the function element z along the trace leads back to the initial element. Because of the assumed simple connectivity of $\mathfrak{F}$ this covering surface must be one sheeted. Thus the proof of the monodromy theorem is already complete.

§ 9. Differentials and line integrals. Homology

The *Cauchy integral theorem* is a special case of the monodromy theorem. To formulate the theorem generally we must discuss "differentials" on a Riemann surface. While a "function" is characterized by the fact that it has a definite value at each point of its domain of definition, a *differential* dz has no intrinsic value at a point $\mathfrak{p}$; rather, dz has a definite (complex) value, $(dz)_t^{\mathfrak{p}}$, only in relation to the differential dt of any local parameter t at $\mathfrak{p}$. If t and τ are local parameters at the same point $\mathfrak{p}$, then

$$(9.1) \qquad (dz)_t^{\mathfrak{p}} = (dz)_\tau^{\mathfrak{p}} \left(\frac{d\tau}{dt}\right)_{t=0}$$

must always hold. A regular analytic function z in a domain $\mathfrak{G}$ has a differential dz for which

$$(dz)_t^{\mathfrak{p}} = \left(\frac{dz}{dt}\right)_{t=0},$$

where $\mathfrak{p}$ is any point of $\mathfrak{G}$ and t is any local parameter at $\mathfrak{p}$. Also, a harmonic function u gives rise to a differential dw, according to the formula

$$(9.2) \qquad (dw)_t^{\mathfrak{p}} = \left(\frac{\partial u}{\partial x} - i\frac{\partial u}{\partial y}\right)_{t=0} \qquad (t = x + iy).$$

Multiplication of a differential dz by a function f yields a new differential dZ:

$$(dZ)_t^{\mathfrak{p}} = f(\mathfrak{p})(dz)_t^{\mathfrak{p}}.$$

If at a point $\mathfrak{p}$, $(dz)_t^{\mathfrak{p}} = 0$ for *one* local parameter t at $\mathfrak{p}$, then this is the case for *every* local parameter at $\mathfrak{p}$; $\mathfrak{p}$ is then a *zero* of dz. If dZ and dz are two differentials defined in the same domain $\mathfrak{G}$ and if dz has no zeros, then $f = dZ/dz$ is a function in $\mathfrak{G}$; for

$$\frac{(dZ)_t^{\mathfrak{p}}}{(dz)_t^{\mathfrak{p}}} = f(\mathfrak{p})$$

is independent of the choice of the local parameter t.

Let t be a local parameter at the point $\mathfrak{p}_0$ and let $|t| < r$ be an associated neighborhood corresponding to the domain $\mathfrak{u}_r$ on $\mathfrak{F}$, which contains $\mathfrak{p}_0$ as an interior point. Suppose that r is small enough so that dz is defined in $\mathfrak{u}_r$.

At each point $\mathfrak{p}$ of $\mathfrak{u}_r$ let the value of dz relative to the local parameter $t - t(\mathfrak{p})$ be denoted by dz/dt. If r may be chosen small enough so that dz/dt is a regular analytic function represented by a power series in t in $|t| < r$, then dz is called *regular analytic* at $\mathfrak{p}_0$. Then it is also regular analytic at all points of a sufficiently small neighborhood of $\mathfrak{p}_0$, and in changing to another local parameter τ at $\mathfrak{p}_0$ an equation of the type (9.1) is valid not only at the origin $t = 0$ but in a whole neighborhood $|t| < r$:

$$\frac{dz}{dt} = \frac{dz}{d\tau}\frac{d\tau}{dt}.$$

Thus the concept of a regular analytic differential at a point $\mathfrak{p}_0$ is independent of the choice of local parameter.

If dz/dt has a zero of order m at $t = 0$, then we also say that the differential dz has a *zero of order m* (or is of order m) at $\mathfrak{p}_0$. If dz/dt has a pole of order n (then dz is defined in a neighborhood of $\mathfrak{p}_0$ except at $\mathfrak{p}_0$ itself), then we say that dz has a *pole of order n* (or is of order $-n$) at $\mathfrak{p}_0$. These orders are independent of the choice of the local parameter t. The same is true of the coefficient A_{-1} in the development

$$\frac{dz}{dt} = A_{-n}t^{-n} + \cdots + A_{-1}t^{-1} + A_0 + A_1 t + A_2 t^2 + \cdots.$$

This is most easily seen from the equation

$$\int\limits_{\kappa} \frac{dz}{dt}dt = 2\pi i\, A_{-1},$$

where κ is any curve in the t-plane which lies in a sufficiently small neighborhood $|t| < r$ and winds around the origin $t = 0$ once in the positive sense (without passing through the origin). When one formulates the conditions on κ this way, instead of taking κ to be a circle, it is clear that the left-hand side of our equation is in fact not affected by the choice of the local parameter t. Because of this equation I prefer to call $2\pi i\, A_{-1}$, not A_{-1} (as is the usual custom), the *residue* of dz at the point $\mathfrak{p}_0$.[21] If one wishes to prove the same fact by algebraic manipulation, then one writes the new local parameter τ in the form $\tau = ct\,\eta(t)$, where $c \neq 0$ and $\eta(t)$ has the power series development $1 + \gamma_1 t + \cdots$. One must show that the coefficient of t^{-1} in the development

---

[21]) One must always bear in mind that a residue is something associated with a *differential*, not a *function*.

of $\tau^{-(\nu+1)}\, d\tau/dt$ in powers of t has the value 1 for $\nu = 0$ and the value 0 for $\nu = 1, 2, \ldots$. The first follows from the equation

$$\tau^{-1}\frac{d\tau}{dt} = t^{-1} + \frac{\eta'}{\eta} \qquad \left(\eta' = \frac{d\eta}{dt}\right).$$

The second follows from the fact that for $\nu \geq 1$, $-\nu c^{\nu}\tau^{-(\nu+1)}\, d\tau/dt$ is the derivative with respect to t of

$$c^{\nu}\tau^{-\nu} = t^{-\nu}\eta^{-\nu}(t) = t^{-\nu} + \cdots$$

and hence the term in t^{-1} is missing.

This last remark also shows that the differential of a function which is regular except for poles is itself regular analytic except for poles and has nowhere a nonzero residue. The converse of this theorem, as far as it is correct, is the content of the Cauchy integral theorem in its general formulation.

If the differential dz is regular except for poles in a simply connected domain $\mathfrak{G}$ *and if dz has nowhere a residue* $\neq 0$, *then there exists a uniform function in* $\mathfrak{G}$, *regular except for poles, whose differential coincides with the given dz throughout* $\mathfrak{G}$.

Proof. Let $\mathfrak{p}_0$ be any point of $\mathfrak{G}$ and let

$$\frac{dz}{dt} = A_{-n}t^{-n} + \cdots + A_{-2}t^{-2} + A_0 + A_1 t + \cdots,$$

for sufficiently small t, where t is a local parameter at $\mathfrak{p}_0$. But this is the derivative of the function element

$$z = -\frac{A_{-n}}{n-1}t^{-(n-1)} - \cdots - \frac{A_{-2}}{1}t^{-1} + B + A_0 t + A_1\frac{t^2}{2} + \cdots,$$

which contains an arbitrary additive constant B. Thus the problem is solved locally at each point. If we start at a fixed point $\mathfrak{p}_0$ of $\mathfrak{G}$, at which dz is regular, then it follows that we can continue the associated function element (normalized by $B = 0$) at $\mathfrak{p}_0$,

$$z = A_0 t + A_1\frac{t^2}{2} + \cdots,$$

analytically along arbitrary paths γ in $\mathfrak{G}$, starting at $\mathfrak{p}_0$, without meeting any critical points except poles. (With the aid of this analytic continuation we define the integral $\int_\gamma dz$.) The existence of a uniform function of the desired sort follows from the monodromy theorem.

We switch from Riemann surfaces to *arbitrary smooth surfaces*. Here the linear differential form corresponds to the differential. If two continuous real-valued functions $f_x(x, y)$ and $f_y(x, y)$ are given in a domain $\mathfrak{G}$ of the xy-plane, then one can regard these as the components of a vector or of a *linear differential form*

$$df = f_x(x, y)\,dx + f_y(x, y)\,dy. \tag{9.3}$$

If $\mathfrak{G}$ is mapped onto a domain $\mathfrak{G}^*$ by a continuously differentiable topological transformation $(x, y) \to (x^*, y^*)$, then (9.3) is to remain invariant under this transformation. That is, the components f_x^* and f_y^* of the same vector relative to (x^*, y^*) are to be determined by the equations

$$f_x^* \frac{\partial x^*}{\partial x} + f_y^* \frac{\partial y^*}{\partial x} = f_x, \qquad f_x^* \frac{\partial x^*}{\partial y} + f_y^* \frac{\partial y^*}{\partial y} = f_y. \tag{9.4}$$

It is clear how the concept is to be defined locally at any point $\mathfrak{o}$ of the smooth surface $\mathfrak{F}$. Relative to an admissible local coordinate system (x, y) at $\mathfrak{o}$, a *vector df* has two real components f_x and f_y at $\mathfrak{o}$. The components in two admissible local coordinate systems (x, y) and (x^*, y^*) are related by the equations (9.4), where now naturally the four derivatives are to be evaluated at the origin:

$$f_x = a_{xx} f_x^* + a_{yx} f_y^*, \qquad f_y = a_{xy} f_x^* + a_{yy} f_y^*.$$

This definition is based on the fact that with a local continuously differentiable topological map $(x, y) \to (x^*, y^*)$ at the origin O there is associated a linear transformation of the "differentials" (dx, dy)

$$dx^* = a_{xx}\,dx + a_{xy}\,dy, \qquad dy^* = a_{yx}\,dx + a_{yy}\,dy$$

such that composition of the maps is mirrored in the composition of the corresponding linear transformations. We will assume that the vector df is given not only at the point $\mathfrak{o}$ of the surface but is defined and continuous in a neighborhood. This is to say that there exists a square neighborhood $|x| < \varepsilon$, $|y| < \varepsilon$ "*belonging*" to the admissible local coordinate system (x,y), whose image on the surface is $\mathfrak{U}_\varepsilon$, with the following property. If (x_0, y_0) is any point of this neighborhood, then the components f_x and f_y at $\mathfrak{p}_0$, relative to the admissible local coordinates $(x - x_0, y - y_0)$ at $\mathfrak{p}_0$, are continuous functions of x_0, y_0 in the square. (Here it is somewhat more convenient to use square or rectangular neighborhoods instead of circular ones; one can of course stick to circular neighborhoods.) If (x^*, y^*) is another admissible local coordinate system at $\mathfrak{o}$, then one can choose ε so small that x^* and y^* are defined and continuously differentiable functions of x, y in the square and have a nonvanishing functional determinant there; then for each point (x_0, y_0) of the square, $x^* - x_0^*, y^* - y_0^*$ is also an admissible local coordinate system at $\mathfrak{p}_0$. The equations (9.4) between the components of df in one coordinate system and in another hold at all points $\mathfrak{p}$ of $\mathfrak{U}_\varepsilon$.

A complex differential $dz = \alpha \, dt$, given in terms of a local parameter $t = x + iy$ at a point $\mathfrak{o}$ of a *Riemann* surface, gives rise to two real linear differential forms $\mathfrak{R}(dz)$, $\mathfrak{J}(dz)$. Namely, if one writes the complex number α, separated into real and imaginary parts, in the form $\alpha = a - ib$, then

$$\alpha \, dt = (a \, dx + b \, dy) + i(-b \, dx + a \, dy).$$

The Cauchy-Riemann differential equations show that for every linear differential form $a \, dx + b \, dy$ defined at the origin there is another (conjugate differential form) $-b \, dx + a \, dy$ and that the relation between these two is invariant under conformal transformations.

We return to the linear differential forms on an arbitrary smooth surface $\mathfrak{F}$. If $\mathfrak{f}(\mathfrak{p})$ is defined and everywhere continuously differentiable in a domain $\mathfrak{G}$ on $\mathfrak{F}$ and if it can be expressed as the function $f(x, y)$ in a neighborhood of $\mathfrak{p}_0$, where (x, y) is an admissible local coordinate system at $\mathfrak{p}_0$, then the values of the derivatives $f_x = \partial f/\partial x$, $f_y = \partial f/\partial y$ at the origin determine a vector $d\mathfrak{f}$ at $\mathfrak{p}_0$ which is independent of the choice of the local coordinate system. Thus one obtains from $\mathfrak{f}$ a continuous vector field $d\mathfrak{f}$ in $\mathfrak{G}$ for which the notation grad $\mathfrak{f}$ has become customary. The vector fields which arise in this way from continuously differentiable functions $\mathfrak{f}$ of points are called *exact.*[22]

With analytic differentials on Riemann surfaces in mind we direct our attention to those continuous linear differential forms df on a smooth surface $\mathfrak{F}$ (or a subdomain of $\mathfrak{F}$) which are *locally exact.* That is, we assume that each point $\mathfrak{p}$ of the domain of definition has a neighborhood in which there is a continuously differentiable function f of points such that $df = \text{grad} f$ (in coordinates, $f_x = \partial f/\partial x$, $f_y = \partial f/\partial y$) in that neighborhood. Such linear differential forms may be called *closed differentials.* In the case of Riemann surfaces we treated the step from an analytic differential dz to its integral $\int_\gamma dz$ along a curve γ rather cursorily; here we shall discuss the integration of closed differentials on smooth surfaces with painful precision. The integral of a continuous linear differential form along a path γ can be defined in the

[22]) The use of this term varies in the literature; frequently "exact" is used in the sense of "closed."

familiar way only when γ is a continuously differentiable (or more generally, a rectifiable) curve. If, however, we have an *exact* vector field grad $\mathfrak{f}$, then the integral has the same value, $\mathfrak{f}(\mathfrak{b}) - \mathfrak{f}(\mathfrak{a})$, for all rectifiable curves joining $\mathfrak{a}$ to $\mathfrak{b}$. Thus this value may be assigned as the value of the integral along any *arbitrary* continuous curve γ from $\mathfrak{a}$ to $\mathfrak{b}$, without demanding that γ be rectifiable. By partitioning the path γ into small pieces one extends this argument from *exact* to *closed* (that is, everywhere locally exact) differentials. The detailed development of this idea goes as follows.

If f is a continuously differentiable function in the square $|x| < a$, $|y| < a$, both of whose derivatives $\partial f/\partial x$, $\partial f/\partial y$ vanish identically, then f is constant in the whole square. This familiar fact from the elements of analysis carries over to arbitrary open sets $\mathfrak{G}$ in the xy-plane as follows.

Lemma. If $f(x, y)$ is a given continuously differentiable function in $\mathfrak{G}$ such that both $\partial f/\partial x$ and $\partial f/\partial y$ vanish identically in $\mathfrak{G}$ and if $\mathfrak{a}$ and $\mathfrak{b}$ are two points of $\mathfrak{G}$, joined by a curve γ in $\mathfrak{G}$, then $f(\mathfrak{a}) = f(\mathfrak{b})$.

The proof follows by enclosing each point of the curve in a square neighborhood and using an associated standard subdivision of the curve. Then one finds from the introductory remark that f has the same value at any two adjacent partition points.

Let now the differential df be defined and closed on $\mathfrak{F}$ and let $\gamma : \mathfrak{p} = \mathfrak{p}(\lambda)$ be a curve on $\mathfrak{F}$. For each point λ_0 of the curve, $\mathfrak{p}_0 = \mathfrak{p}(\lambda_0)$, we determine a smooth local coordinate system (x, y) and an associated square neighborhood $|x| < a$, $|y| < a$ in which the components f_x and f_y of $df = f_x\,dx + f_y\,dy$ are the derivatives of $\partial f/\partial x$ and $\partial f/\partial y$ of a function $f(x, y)$ which is continuously differentiable in the square. Let $\mathfrak{u}(\mathfrak{p}_0)$ be the image of the square on the surface $\mathfrak{F}$. One can split γ into finitely many subarcs $\gamma_i = (\mathfrak{a}_{i-1}\,\mathfrak{a}_i)$ $(i = 1, \ldots, n)$ and choose a point $\mathfrak{p}_i = \mathfrak{p}(\lambda_i)$ on each γ_i such that γ_i is contained completely in $\mathfrak{u}_i = \mathfrak{u}(\mathfrak{p}_i)$. The notations x_i, y_i, f_i are clear without more ado. Now we define

$$\int_\gamma df = \sum_{i=1}^{n} \Delta_i f_i, \qquad \text{where} \qquad \Delta_i f_i = f_i(\mathfrak{a}_i) - f_i(\mathfrak{a}_{i-1}).$$

It must be proved that the result depends only on df and γ, not on the construction used.

If one carries out the process described in a second way by splitting γ into subarcs γ_j^* by partition points $\mathfrak{a}_j^*$, then one obtains $\sum_j \Delta_j^* f_j^*$ as the new value of $\int_\gamma df$. The meaning of x_j^*, y_j^*, and f_j^* is clear. By superposition of the two partitions one obtains a finer partition of γ into arcs γ', each of which is the intersection of a γ_i with a γ_j^*. For the moment set $x = x_i$, $y = y_i$; $x^* = x_j^*$, $y^* = y_j'$; $f = f_i$, $f^* = f_j^*$; $\mathfrak{u} = \mathfrak{u}(\mathfrak{p}_i)$, $\mathfrak{u}^* = \mathfrak{u}^*(\mathfrak{p}_j^*)$. The small arc γ' lies in the intersection $\mathfrak{u} \cap \mathfrak{u}^*$. In that intersection x^* and y^* are continuously differentiable functions of x and y with a nonvanishing functional determinant; the components $f_x = \partial f/\partial x$, $f_y = \partial f/\partial y$ and $f_x^* = \partial f^*/\partial x^*$, $f_y^* = \partial f^*/\partial y^*$ are connected by the equations (9.4) and hence

$$\frac{\partial f^*}{\partial x} = \frac{\partial f}{\partial x}, \qquad \frac{\partial f^*}{\partial y} = \frac{\partial f}{\partial y}$$

in $\mathfrak{u} \cap \mathfrak{u}^*$. If γ' runs from the point $\mathfrak{a}$ to the point $\mathfrak{a}'$, then it follows from the Lemma that the increment of f^* along γ', that is, $f^*(\mathfrak{a}') - f^*(\mathfrak{a})$, is equal to the increment of f. But the increment $\Delta_i f_i$ of f_i on γ_i is equal to the sum of its increments on the finer arcs γ' into which γ_i is split; the increment $\Delta_j^* f_j^*$ along γ_j^* is equal to the sum of the increments of f_j^* on the finer arcs γ' into which γ_j^* is split. Thus, if γ_{ij} denotes the intersection of γ_i and γ_j^*,

$$\sum_i \Delta_i f_i = \sum_{i,j} \Delta_{ij} f_i = \sum_{i,j} \Delta_{ij} f_j^* = \sum_j \Delta_j^* f_j^*.$$

A curve γ_1 from $\mathfrak{a}$ to $\mathfrak{b}$ and a curve γ_2 from $\mathfrak{b}$ to $\mathfrak{c}$ can be joined to form a curve $\gamma = \gamma_1 + \gamma_2$ which runs from $\mathfrak{a}$ through $\mathfrak{b}$ to $\mathfrak{c}$. Then always

$$\int_{\gamma_1} df + \int_{\gamma_2} df = \int_{\gamma} df.$$

For a closed path γ the integral $\int_\gamma df$ is independent of the point at which one starts to trace the path. The exact differentials can now be characterized as the closed differentials for which the *integral* $\int_\gamma df$ *vanishes along every closed path* γ.

Here it is possible, not only to ascend from Riemann surfaces to arbitrary smooth surfaces, but to extend the concepts to the full generality of topological surfaces. The resulting concept is that of *integral function.* A *function of curves, F,* on the surface $\mathfrak{F}$ is defined if with each curve γ on $\mathfrak{F}$ a (real) number $F(\gamma)$ is associated. The function of curves is *linear* if $F(\gamma_1 + \gamma_2) = F(\gamma_1) + F(\gamma_2)$ holds. F is called *exact* or *cohomologous to zero,* $F \sim 0$, if $F(\gamma) = 0$ for every *closed* curve γ. In this case there exists a function $\mathfrak{f}(\mathfrak{p})$ of points on the surface such that F is the increment function of $\mathfrak{f}$; that is, for any curve γ from $\mathfrak{a}$ to $\mathfrak{b}$ the value of $F(\gamma)$ is the increment $\mathfrak{f}(\mathfrak{b}) - \mathfrak{f}(\mathfrak{a})$ of $\mathfrak{f}$. I have called a linear function of curves an *integral function* (the concept was introduced in the first edition of this book) if each point of the surface has a neighborhood such that $F(\gamma) = 0$ for every closed curve γ in this neighborhood. This is the topological generalization of continuous closed differentials.

An integral function F on a simply connected surface $\mathfrak{F}$ is necessarily exact. For choose a fixed point $\mathfrak{o}$ on $\mathfrak{F}$ and with each curve γ from $\mathfrak{o}$ to a point $\mathfrak{p}$ of $\mathfrak{F}$ associate a point $\bar{\mathfrak{p}}$ over $\mathfrak{p}$ so that two curves γ_1 and γ_2 from $\mathfrak{o}$ to $\mathfrak{p}$ determine the same point $\bar{\mathfrak{p}}$ if and only if $F(\gamma_1) = F(\gamma_2)$. The result is a perfect covering surface over $\mathfrak{F}$. If $\mathfrak{F}$ is simply connected, this covering must be one sheeted, and that says that $F(\gamma_1) = F(\gamma_2)$ for any two curves γ_1 and γ_2 from $\mathfrak{o}$ to $\mathfrak{p}$. If one denotes this value by $\mathfrak{f}(\mathfrak{p})$, then F is the increment function of $\mathfrak{f}$.

More generally, we can say that the covering surface $\bar{\mathfrak{F}}$ which we have just constructed on the basis of a given integral function F, on which a curve $\bar{\gamma}$ with closed trace γ is closed if and only if $F(\gamma) = 0$, is a normal covering surface and *its cover transformations S form an Abelian = commutative group* Γ. In fact, as on page 58, let $\bar{\mathfrak{o}}$ be a point of $\bar{\mathfrak{F}}$ over the point $\mathfrak{o}$ on $\mathfrak{F}$ and let α and γ be curves on $\mathfrak{F}$ which start and end at $\mathfrak{o}$. The cover transformation TS composed of $S = S_\alpha$ and $T = S_\gamma$ carries $\bar{\mathfrak{o}}$ into the point obtained by starting at $\bar{\mathfrak{o}}$ and following over the path $\alpha + \gamma$. But since $F(\alpha + \gamma) = F(\gamma + \alpha)$, we reach the same point with the path $\gamma + \alpha$, and hence $TS = ST$.

The concept of the *residue* may be extended to arbitrary closed differentials on smooth surfaces, and even to arbitrary integral functions on surfaces. Let a neighborhood $\mathfrak{U}$ of the point $\mathfrak{o}$ on $\mathfrak{F}$ be mapped topologically onto the interior E of the unit circle in the xy-plane so that $\mathfrak{o}$ corresponds to the origin O. Polar coordinates are introduced by the equations

$$z = x + iy = r \cdot \mathrm{ex}\,\phi, \qquad x = r \cdot \mathrm{co}\,\phi, \qquad y = r \cdot \mathrm{si}\,\phi,$$

where $r \geq 0$ and we have set

$$\mathrm{ex}\,\phi = e^{2\pi i \phi}, \qquad \mathrm{co}\,\phi = \cos 2\pi\phi, \qquad \mathrm{si}\,\phi = \sin 2\pi\phi.$$

If the origin O is punched out of E ("delected disc" $E^{\cdot}$), then one obtains the infinite parallel strip $0 < r < 1$ in the $r\phi$-plane as a covering surface over $E^{\cdot}$ with branch point O: the points (r, ϕ) and $(r, \phi + n)$ coincide on $E^{\cdot}$ if n is any integer. A closed curve γ in the deleted neighborhood $\mathfrak{U}^{\cdot}$ appears in the (r, ϕ)-image as a curve from the point (r, ϕ) to a point $(r, \phi + n)$. The integer n indicates how many times the image curve in $E^{\cdot}$ goes around the origin. An integral function F defined in $\mathfrak{U}^{\cdot}$ carries over into an integral function in the parallel strip. Since the strip is simply connected, $F(\gamma) = 0$ for every curve γ which is closed, not only in $E^{\cdot}$ but also in the strip. Thus for all closed curves γ in $\mathfrak{U}^{\cdot}$, which have the same winding number n, $F(\gamma)$ has the same value $A^{(n)}$. For if one, γ_0, runs from (r_0, ϕ_0) to $(r_0, \phi_0 + n)$ in the strip, and another, γ_1, runs from (r_1, ϕ_1) to $(r_1, \phi_1 + n)$, then join the initial points (r_0, ϕ_0) and (r_1, ϕ_1) with a linear segment and join the terminal points with a parallel segment. Applying $F(\gamma) = 0$ to the closed curve consisting of γ_0, γ_1, and these two segments we obtain $F(\gamma_0) = F(\gamma_1)$. Also it is clear that $A^{(n)} = nA^{(1)}$, where $A^{(1)} = A$ is the value of F for a circle in E with center O; for the circle traced n times winds about the point O n times. A is called the *residue* of F at $\mathfrak{o}$.

Now we consider integral functions which are defined for all paths γ on the surface $\mathfrak{F}$. They form a linear space; for multiplication of an integral function by a real constant results in another integral function, and similarly with the addition of two integral functions. The integral functions $F_1, \ldots, F_l$ are *linearly independent* ("in the sense of cohomology") if there does not exist a set of real numbers $c_1, \ldots, c_l$, not all zero, such that

$$c_1 F_1 + \cdots + c_l F_l \sim 0.$$

If any $h + 1$ integral functions on $\mathfrak{F}$ are linearly dependent and if there exists a set of h linearly independent integral functions, then h is the dimension of the linear family of integral functions on $\mathfrak{F}$; we call h the *degree (degree of connectivity)* of the surface $\mathfrak{F}$. Any h linearly independent integral functions $F_1, \ldots, F_h$ may be used as a *basis* for the linear family of integral functions in the sense that any integral function F is cohomologous to one and only one linear combination of $F_1, \ldots, F_h$:

$$F \sim c_1 F_1 + \cdots + c_h F_h, \qquad \text{i.e.,} \qquad F - c_1 F_1 - \cdots - c_h F_h \sim 0.$$

Naturally it may happen that the degree of a surface is infinite. A *homology*

$$c_1 \gamma_1 + \cdots + c_l \gamma_l \simeq 0 \tag{9.5}$$

with real coefficients c_i between closed paths γ_i means that

$$c_1 F(\gamma_1) + \cdots + c_l F(\gamma_l) = 0$$

for each integral function F on $\mathfrak{F}$. If one introduces formally such linear combinations of closed paths

$$\sigma = c_1 \gamma_1 + \cdots + c_l \gamma_l \tag{9.6}$$

as "*streams*" and if one does not distinguish between homologous streams, then the streams form a linear space which is dual to that of the integral functions if the inner product of the integral function F and the stream (9.6) is the number $c_1 F(\gamma_1) + \cdots + c_l F(\gamma_l)$. Thus on a surface of degree h, any $h + 1$ closed paths satisfy a homology of the type (9.5) with not all coefficients c_i vanishing; on the other hand, there exist h closed paths which are linearly independent in the sense of homology and may be used as a basis of the closed paths.[23]

For each curve γ one can form the curve traced in the opposite sense: $-\gamma$. If γ and γ' are two closed curves, then one can form a closed curve $\gamma + \gamma'$ as follows: trace γ, starting at a point $\mathfrak{a}$, trace a path β from $\mathfrak{a}$ to a point $\mathfrak{a}'$ of γ', then trace γ' from $\mathfrak{a}'$ to $\mathfrak{a}'$, and finally trace $-\beta$ from $\mathfrak{a}'$ back to $\mathfrak{a}$. For every integral function F the relations

$$F(-\gamma) = -F(\gamma), \qquad F(\gamma + \gamma') = F(\gamma) + F(\gamma')$$

are valid for any closed paths γ and γ'. Thus one can also interpret any linear combination $n_1 \gamma_1 + \cdots + n_l \gamma_l$ of closed paths γ_i with *integral* coefficients n_i as a closed path instead of as a stream.

If, in a linear space $\mathfrak{S}$ of integral functions, one considers a linear subspace $\mathfrak{S}'$, then the degree of the family can decrease at most; it remains the same if each integral function F in $\mathfrak{S}$ is cohomologous to an F' in $\mathfrak{S}'$: $F - F' \sim 0$. In particular, this fact is applied in the case of a smooth surface $\mathfrak{F}$ when, among all integral functions, we restrict ourselves to those which arise from integration of closed differentials. Similarly on a Riemann surface if, from the linear space of closed differentials, we single out the family of differentials which are everywhere harmonic (which are the real parts of analytic differentials). In both cases it turns out that, at least for closed surfaces, the degree does *not* decrease; the proof for the second case will be given in Chapter II. In the study of harmonic (or analytic) differentials on Riemann surfaces, the question arises as to whether we should depend on a consideration of closed differentials on arbitrary smooth surfaces or integral functions on surfaces. Generalization entails facilitation in that it simplifies the situation with which one has to deal; it is fruitful if essential features of the phenomena are not erased. Led by this point of view, we shall commence our study of functions and integrals on Riemann surfaces with a study of arbitrary *smooth* surfaces. Going back to *topological* surfaces would hardly yield further simplifications; and the smooth surfaces have the great advantage that one can use the linearizing processes of the differential calculus on them; for surfaces in

---

[23]) One may well say that this definition of the degree exposes the real kernel of the method used by Weierstrass, Hensel and Landsberg, among others, in the theory of algebraic functions: to investigate the behavior of *integrals* first, and to draw conclusions therefrom about the *paths* of integration. The passage from closed paths (or streams) to the dual linear space of integral functions has been developed extensively and systematized during the last decade as cohomology theory in topology. In that case manifolds and paths of arbitrary dimension are considered.

general the intrinsically more complicated procedure of *simplical approximation* is indicated. Thus from now on we use as the definition of the *degree* h of a smooth surface $\mathfrak{F}$, the *maximum number of linearly independent closed differentials on* $\mathfrak{F}$. In particular $\mathfrak{F}$ is called *completely planar* (schlicht) if $h = 0$, that is, if every closed differential on $\mathfrak{F}$ is exact. We have noted above that if $\mathfrak{F}$ is simply connected, then $\mathfrak{F}$ is completely planar. The *homology* (9.5) now means for us that the equation

$$c_1 \int_{\gamma_1} df + \cdots + c_l \int_{\gamma_l} df = 0 \tag{9.7}$$

holds for every closed differential df on $\mathfrak{F}$.

A closed differential df is called *compact* if there exists a compact set on $\mathfrak{F}$ outside of which $df = 0$. If the dimension of the linear space of compact closed differentials is finite, it will be called the *weak degree of connectivity* of $\mathfrak{F}$. It cannot exceed the degree of connectivity h.

The *weak homology*

$$c_1\gamma_1 + \cdots + c_l\gamma_l \sim 0$$

means that the relation (9.7) is valid for every *compact* closed differential df on $\mathfrak{F}$. For closed surfaces there is of course no distinction between homology and weak homology. But, for example, the interior of a circular annulus in the plane has degree of connectivity 1 and weak degree of connectivity 0. For if the annulus $a < r < b$, then any circle r = constant in the annulus is not homologous to 0 but is weakly homologous to 0. A surface on which all compact closed differentials are exact, that is, a surface whose weak degree of connectivity is 0, is called *planar* (schlichtartig). There is the following noteworthy theorem.

Every subdomain $\mathfrak{G}$ of a planar surface is also planar.

Proof. If df is a compact closed differential in $\mathfrak{G}$, then df vanishes outside a compact subset of $\mathfrak{G}$. If we set $df = 0$ at every point of $\mathfrak{F}$ outside that compact subset of $\mathfrak{G}$, then we obtain a compact closed differential on $\mathfrak{F}$. Since $\mathfrak{F}$ is planar, df must be exact on $\mathfrak{F}$ and, a fortiori, exact on $\mathfrak{G}$.

One can construct a normal perfect covering surface $\hat{\mathfrak{F}}$ over $\mathfrak{F}$ with the following property. A point in $\hat{\mathfrak{F}}$ whose trace in $\mathfrak{F}$ follows a closed curve γ describes a closed curve on $\hat{\mathfrak{F}}$ if and only if γ is weakly homologous to zero. We will call this surface the *class surface* of $\mathfrak{F}$, for it is analogous to what is called the class field in the theory of algebraic number fields. To every closed path γ on $\mathfrak{F}$ there corresponds a definite cover transformation $S = S_\gamma$ of $\hat{\mathfrak{F}}$ which is unaltered if γ is replaced by another path weakly homologous to γ. If one follows over the path γ on $\hat{\mathfrak{F}}$, starting at a point $\hat{\mathfrak{p}}$, it leads to $\hat{\mathfrak{p}}S$.

The group of cover transformations is Abelian, corresponding to the rule $S_{\alpha+\gamma} = S_\alpha S_\gamma$ for any two closed paths α and γ, and is isomorphic to the additive group of closed paths if weakly homologous paths are identified.

Let df be a continuous closed differential on the surface obtained by deleting the finite set of points $\mathfrak{a}_1, \ldots, \mathfrak{a}_n$ from the surface $\mathfrak{F}$; let the residues of df be zero at each of these points. If α is a closed path which avoids the singular points $\mathfrak{a}_i$ and if $\alpha \simeq 0$, then

$$\int_\alpha df = 0. \tag{9.8}$$

The point of this theorem is that one can extend the formula (9.8) from differentials df which are everywhere continuous to differentials which have singularities without residues. If df is compact, then it is enough to assume that α is weakly homologous to zero. We supply the proof for a single singularity $\mathfrak{a}_1 = \mathfrak{o}$; the extension of the proof to several singularities is trivial. We pick a disc $\mathfrak{k} : r \leq a$, in an admissible local coordinate system at $\mathfrak{o}$. Since the residue is zero, df is the differential of a continuously differentiable function f in $\mathfrak{k}^\cdot$ (that is, the disc $\mathfrak{k}$ with the origin deleted). We choose a smaller disc $\mathfrak{k}_1 : r \leq a_1$, and a continuously differentiable function $\lambda(r)$ which is identically 1 for $r \leq a_1$, is identically 0 for $r \geq a$, and which decreases from 1 to 0 in the interval $a_1 \leq r \leq a$. We call λ a *smoothing function* in the annulus (a_1, a). Then $\lambda f = \eta$ is a continuously differentiable function on the whole deleted surface $\mathfrak{F}^\cdot$ which vanishes outside $\mathfrak{k}$ and coincides with f in $\mathfrak{k}_1^\cdot$. The "smoothed" differential $df' = df - d\eta$, which differs from df only in $\mathfrak{k}$, is also continuous at the point $\mathfrak{o}$; for it vanishes in a whole neighborhood of $\mathfrak{o}$. If $\mathfrak{k}$ is chosen so small that the curve α does not penetrate $\mathfrak{k}$, then clearly

$$\int_\alpha df = \int_\alpha df', \tag{9.9}$$

and hence $\alpha \simeq 0$ implies (9.8). In truth, (9.9) is valid for any closed curve α not passing through $\mathfrak{o}$, since $d\eta$ is an exact differential on $\mathfrak{F}^\cdot$; this remark will be of importance for us later.

The degree of connectivity of a compact surface is finite.

The proof of this fundamental theorem utilizes the discs $\mathfrak{R}_i$ and the bridges $\mathfrak{B}^{(\rho)}$ which were used in § 8 to prove the axiom of countability for covering surfaces. But here we have the simplification that $\mathfrak{F}$ is covered with *finitely many* $\mathfrak{R}_i$; for $\mathfrak{F}$ is assumed to be compact. If F is any integral function on $\mathfrak{F}$, then without altering the value $F(\gamma)$ any closed path γ may be replaced by a closed path γ' which consists of pieces $\gamma_{ij}^{(\rho)} = \mathfrak{a}_i \mathfrak{c}^{(\rho)} \mathfrak{a}_j$. Here $\gamma_{ij}^{(\rho)}$ consists of one piece $\mathfrak{a}_i \mathfrak{c}^{(\rho)}$ in $\mathfrak{U}_i$ which joins the centers $\mathfrak{a}_i$ and $\mathfrak{c}^{(\rho)}$ of $\mathfrak{U}_i$ and the bridge $\mathfrak{B}^{(\rho)}$, and of a second piece $\mathfrak{c}^{(\rho)}\mathfrak{a}_j$ in $\mathfrak{U}_j$. Certainly then the integral function

F is cohomologous to zero if the $R = \sum_{i<j} r_{ij}$ numbers $f_{ij}^{(\rho)} = F(\gamma_{ij}^{(\rho)})$ all vanish; hence there cannot exist more than R integral functions which are linearly independent in the sense of cohomology.

§ 10. Densities and surface integrals. The residue theorem

Let $\mathfrak{F}$ be an oriented smooth surface, $\mathfrak{o}$ a point on it, (x, y) an admissible local coordinate system at $\mathfrak{o}$, and let $\mathfrak{B}$ given by

$$(10.1) \qquad -a' < x < a', \qquad -b' < y < b' \qquad (a', b' > 0)$$

be a rectangular neighborhood of the origin which corresponds to a domain $\mathfrak{u}$ about $\mathfrak{o}$ on $\mathfrak{F}$. Relative to such a local coordinate system a *density* ψ has a definite real value q at $\mathfrak{o}$ and the values q and q^* in two admissible local coordinate systems (x, y) and (x^*, y^*) are connected by the equation

$$(10.2) \qquad q^* J = q,$$

where J is the (positive) functional determinant of the continuously differentiable transformation $(x, y) \to (x^*, y^*)$ at the origin. If (x_0, y_0) is a point of the rectangle (10.1), corresponding to the point $\mathfrak{p}_0$ of $\mathfrak{u}$, and if the density ψ is defined in all of $\mathfrak{u}$, then ψ has a value $q(x_0, y_0)$ at $\mathfrak{p}_0$ relative to the admissible coordinate system $x - x_0, y - y_0$ at $\mathfrak{p}_0$. We express this fact by the equation

$$\psi(\mathfrak{p}_0) = q(x_0, y_0) \qquad \text{or} \qquad \psi(\mathfrak{p}) = q(x, y).$$

The relation (10.2) remains valid for all points sufficiently close to $\mathfrak{o}$, where J is understood to be the functional determinant

$$\frac{\partial x^*}{\partial x}\frac{\partial y^*}{\partial y} - \frac{\partial x^*}{\partial y}\frac{\partial y^*}{\partial x}.$$

ψ is called *continuous* at the point $\mathfrak{o}$ if $q(x, y)$ is continuous at the origin. The equation (10.2) shows that this condition is independent of the choice of the admissible local coordinate system.

Example of a density. If df and dg are two continuous vector fields in a neighborhood of a point $\mathfrak{o}$ on $\mathfrak{F}$ which, in terms of an admissible local coordinate system, are given by

$$df = f_x\, dx + f_y\, dy, \qquad dg = g_x\, dx + g_y\, dy,$$

then the "*skew product*"

$$(10.3) \qquad q = f_x g_y - f_y g_x = [df, dg]$$

is a continuous density. For the transformations (9.4) for f_x, f_y and g_x, g_y

imply that the right-hand side of (10.3) satisfies the transformation (10.2).

Inside $\mathfrak{u}$, (10.1), we pick a compact *block B*

(10.4) $$-a \leq x \leq a, \qquad -b \leq y \leq b$$

with center $\mathfrak{o}$ (and coordinates x, y); here a and b are any two positive numbers which satisfy the conditions $a < a'$, $b < b'$. To begin with we assume that $\mathfrak{F}$ is closed (compact). I claim: *with each density ψ which is defined and continuous on all of $\mathfrak{F}$ one can associate uniquely an integral $\int\psi$ which satisfies the following two axioms:*

(i) $$\int(\psi_1 + \psi_2) = \int\psi_1 + \int\psi_2 .$$

(ii) *If ψ vanishes outside a block B and if in B, ψ has the value $q(x, y)$ relative to the block coordinates, then $\int\psi$ is equal to the Riemann integral*

$$\int_{-b}^{b} \int_{-a}^{a} q(x, y)\, dx\, dy .$$

Proof. At the time of the first edition of this book one sought to prove this theorem by *partitioning* the surface into nonoverlapping pieces. Today, thanks to a simple artifice introduced by Dieudonné, it is possible to base the proof on a *covering* of the manifold with neighborhoods.

Enclose each point $\mathfrak{p}_0$ of the compact surface $\mathfrak{F}$ in a block $B(\mathfrak{p}_0) = B$ as given by (10.4) for some admissible local coordinate system x, y at $\mathfrak{p}_0$. Choose a positive number $\theta < 1$ and replace B by the "shrunken" block B^θ given by $|x| \leq \theta a$, $|y| \leq \theta b$. We can choose a finite number of points $\mathfrak{p}_i$ $(i = 1, \ldots, n)$ such that the associated shrunken blocks B_i^θ cover $\mathfrak{F}$ completely. Let x_i and y_i be the coordinates belonging to $B_i = B(\mathfrak{p}_i)$. As will be shown shortly, one can construct continuous, and even differentiable, functions μ_i (Dieudonné factors) with the following properties: always $0 \leq \mu_i \leq 1$, μ_i vanishes outside B_i, and the sum $\sum_i \mu_i$ is identically 1. For a given continuous density ψ we take the Dieudonné decomposition

$$\psi = \sum_i \psi_i, \qquad \psi_i = \mu_i \psi ,$$

and compute the integral of ψ_i according to axiom (ii) by means of the coordinates x_i, y_i belonging to B_i. Finally, according to axiom (i), we set

$$\int\psi = \sum_i \int\psi_i .$$

I claim that the functional $\int\psi$ computed thus for any continuous density ψ satisfies both axioms. The first is obvious, the second follows by virtue of the transformation law (10.2). For this says, for a ψ which vanishes outside the block B, that the integral of $\mu_i\psi$ comes out the same whether we compute it with the coordinates $x_i y_i$ belonging to B_i or with the coordinates x, y belonging to B. Also we have seen that the rules (i) and (ii) determine the value $\int\psi$ and hence this value is necessarily independent of the aids used to compute it: the blocks B_i, the coordinates x_i, y_i, and the Dieudonné decomposition by means of the factors μ_i. Besides the rules (i) and (ii), the following law holds:

(iii) $\qquad \int c\psi = c\int\psi \qquad$ for any constant c.

We must still construct the functions μ_i. A continuous function $\lambda(\mathfrak{p})$ which satisfies the inequality $0 \leq \lambda(\mathfrak{p}) \leq 1$ everywhere will be called a *probability function*. For an admissible block B, (10.4), one can easily construct a probability function $\lambda(\mathfrak{p})$ which vanishes outside of B and is identically 1 on the shrunken block B^{θ}. Namely, take a continuous probability function $\lambda(x)$ of the real variable x, which is 1 for $-\theta \leq x \leq \theta$ and which vanishes for $|x| \geq 1$. We can see to it that this function is not only continuous but also continuously differentiable. If we set

$$\lambda(\mathfrak{p}) = \lambda\left(\frac{x}{a}\right)\cdot\lambda\left(\frac{y}{b}\right) \qquad \text{in } B,$$

then we have a function on $\mathfrak{F}$ of the type desired. In this fashion we construct a $\lambda_i(\mathfrak{p})$ for each block B_i. To construct the μ_i from these we use the following elementary operation of probability theory.

If α and β are the probabilities of two statistically independent events, $0 \leq \alpha, \beta \leq 1$, then the probability that one or the other of the two events occur is

$$\alpha \vee \beta = \alpha + \beta - \alpha\beta.$$

Note that

$$\alpha \vee \beta \leq 1, \qquad (\alpha \vee \beta) \geq \alpha \qquad \text{and} \qquad \geq \beta, \qquad \text{a fortiori} \geq 0.$$

The "probability sum" $\alpha \vee \beta$ may be written as an ordinary sum $\alpha + \beta'$, where $\beta' = \beta - \alpha\beta$ lies between 0 and β. The probability sum $\lambda_1 \vee \lambda_2 \vee \cdots$ of the probability functions λ_i associated with the blocks B_i is identically 1 since at every point of the surface at least one $\lambda_i = 1$. This sum may be written as an ordinary sum $\mu_1 + \mu_2 + \cdots$ by starting with $\mu_1 = \lambda_1$ and setting

$$\mu_i = \lambda_i - \lambda_i(\lambda_1 \vee \cdots \vee \lambda_{i-1}).$$

These μ_i possess the required properties and are also continuously differentiable on the whole surface $\mathfrak{F}$.

Instead of assuming that $\mathfrak{F}$ is compact, it is enough if the density ψ is *compact*; that is, ψ vanishes outside a compact subset G of $\mathfrak{F}$. For determine a countable sequence $\mathfrak{p}_i$ ($i = 1, 2, \ldots$) so that every point of the surface is an *interior* point of some shrunken block $B^\theta(\mathfrak{p}_i) = B_i^\theta$. Each point $\mathfrak{p}_0$ of the compact subset G has a neighborhood contained in one of the B_i^θ; by covering G with a finite number of such neighborhoods one obtains a number n such that $\mu_1 + \cdots + \mu_n = 1$ is valid in all of G [and hence all higher μ_i ($i > n$) vanish in G]. Then by the axioms (i) and (ii) we are forced to

$$\int \psi = \sum_{i=1}^{n} \int \mu_i \psi = \sum_{i=1}^{\infty} \int \mu_i \psi . \tag{10.5}$$

For $\mu_i \psi$ vanishes everywhere for $i > n$: in G, where $\mu_i = 0$, and outside G, where $\psi = 0$. Since with this definition the axioms (i) and (ii) hold for all compact continuous densities, then, as above, the value of $\int \psi$ is independent of the construction.

Another situation in which $\int \psi$ may be defined unambiguously on a non-compact surface $\mathfrak{F}$ is the case of densities ψ *which are nonnegative on all of* $\mathfrak{F}$. Of course we must allow ∞ as a value of the integral. In place of axiom (i) one demands here the stronger relation

$$\int (\psi_1 + \psi_2 + \cdots) = \int \psi_1 + \int \psi_2 + \cdots \tag{i*}$$

for *infinite* sums $\psi_1 + \psi_2 + \cdots$ of nonnegative densities ψ_i. From this it follows that $\int \psi$ is to be defined as $\sum_{i=1}^{\infty} \int \mu_i \psi$. Again it turns out that with this definition axioms (i*) and (ii) are satisfied, and hence $\int \psi$ is independent of the construction.

If ψ_1 and ψ_2 are two nonnegative densities whose difference $\psi_2 - \psi_1 = \psi$ is compact, then the equation

$$\int \psi_2 = \int \psi_1 + \int \psi$$

follows from (10.5); this implies that $\int \psi_2$ is finite or infinite according as $\int \psi_1$ is finite or infinite. A continuous density ψ may always be written as the difference $\psi_2 - \psi_1$ of two nonnegative densities ψ_1 and ψ_2. If one sets

$$\left.\begin{aligned} \psi^+ &= \psi \\ \psi^- &= 0 \end{aligned}\right\} \text{if } \psi \geq 0, \qquad \left.\begin{aligned} \psi^+ &= 0 \\ \psi^- &= -\psi \end{aligned}\right\} \text{if } \psi \leq 0,$$

then nonnegative ψ_1 and ψ_2 with the given difference ψ necessarily have the form

$$\psi_2 = \psi^+ + \psi', \qquad \psi_1 = \psi^- + \psi' \qquad \text{with } \psi' \geq 0.$$

Therefore $\int \psi_1$ and $\int \psi_2$ can be finite only if $\int \psi^+$ and $\int \psi^-$ are finite; then $\int \psi$ may be uniquely defined as the difference $\int \psi_2 - \int \psi_1 = \int \psi^+ - \int \psi^-$. Along with the theorem that if a continuous density ψ on $\mathfrak{F}$ vanishes outside a block B, then the integral $\int_{\mathfrak{F}} \psi$ is equal to the Riemann integral over the block

B, $\int_B \psi$, we have the theorem for nonnegative continuous densities $\psi \geq 0$ that $\int_{\mathfrak{F}} \psi \geq \int_B \psi$. It is enough to prove this for the shrunken block B^θ, since the block B can be generated by shrinking a somewhat larger one. If we introduce the probability function λ, used above to construct the Dieudonné factors μ_i, which is identically 1 in B^θ and vanishes outside B ("scarp function"), then, since $\psi \geq \lambda\psi$ everywhere, we obtain the chain of inequalities

$$\int_{\mathfrak{F}} \psi \geq \int_{\mathfrak{F}} \lambda\psi = \int_B \lambda\psi \geq \int_{B^\theta} \psi .$$

Instead of using rectangular blocks one can use discs $K: r \leq a$, relative to a local coordinate system x, y. Then it is appropriate to introduce polar coordinates r, ϕ in place of x, y. Let the density ψ be expressed as $g(x, y)$ in terms of the coordinates x, y. We use the same symbol g also for the function $g(r \text{ co } \phi, r \text{ si } \phi)$ of r and ϕ, although the density ψ in the coordinate system r, ϕ is $2\pi rg$. Instead of the $(z = x + iy)$-disc K, I next consider the annulus $R: b \leq r \leq a$, where $0 < b < a$. Then the following two theorems are valid.

(1) If ψ vanishes outside R, then $\int_{\mathfrak{F}} \psi$ is equal to the Riemann integral

$$\int_R \psi = 2\pi \int_b^a \int_0^1 g \, d\phi \; r \, dr$$

over the annulus.

(2) If ψ is a nonnegative density, then $\int_{\mathfrak{F}} \psi \geq \int_R \psi$.

The proof is in essence the same as for a block. Only in place of the block $b \leq r \leq a$, $0 \leq \phi \leq 1$ in the r, ϕ-plane we have the strip $b \leq r \leq a$ with the prescription that points (r, ϕ) whose ϕ coordinates differ by an integer are to be identified. By passing to the limit $b = 0$ one obtains the second theorem for the disc K. To extend the first theorem to a density vanishing outside K, choose an $\varepsilon < b$ and introduce the quantity $\lambda\psi$, where λ is a smoothing function for the annulus (ε, b). The integrals $\int_{\mathfrak{F}} \lambda\psi$ and $\int_K \lambda\psi$ remain unchanged if $\mathfrak{F}$ is replaced by the square $|x| \leq b$, $|y| \leq b$ and K by the disc $r \leq b$. Both tend to zero with b as rapidly as b^2. By Theorem 1 for the annulus $\varepsilon \leq r \leq a$,

$$\int_{\mathfrak{F}} (1 - \lambda)\psi = \int_K (1 - \lambda)\psi .$$

By passing to the limit $b = 0$ we do indeed obtain $\int_{\mathfrak{F}} \psi = \int_K \psi$.

If $u(x, y)$ is a given continuous function in the disc $K: |z| \leq a$ and if G is a subset of K which possesses a Jordan content (for example, a rectangle or disc), then one can define the integral $\int_G = \int_G u(x, y) \, dx \, dy$ in the familiar Riemannian way. In particular,

$$\int_K = I = 2\pi \int_0^a \int_0^1 u d\phi \; r \, dr.$$

If $u \geq 0$ everywhere, then $\int_G \leq I$. We shall make occasional use of this later.

By means of a smoothing function one can easily define the integral of a continuous density ψ which is *defined and continuous only outside the hole* $\mathfrak{K}: r < a$ *on the surface* $\mathfrak{F}$. (The punched surface $\mathfrak{F} - \mathfrak{K}$ has a *border*.) Choose then a radius b which is somewhat greater than a and a smoothing function λ for the annulus (a, b). To the integral of $(1 - \lambda)\psi$ over $\mathfrak{F}$ add the integral, computed in polar coordinates, of $\lambda\psi$ over the annulus $a \leq r \leq b$; the sum is to regarded as the integral of ψ over the punched surface. It is independent of the choice of b and λ. For two smoothing functions λ and λ' for the annuli (a, b) and (a, b') may be regarded as smoothing functions in the same annulus (a, b) if b is the larger of the two numbers b and b'. The result then follows simply from the fact that the integral of the continuous density $(\lambda - \lambda')\psi$, which vanishes outside the annulus (a, b), has the same value whether one integrates over the whole surface $\mathfrak{F}$ or only over the annulus. This remark is particularly useful if one wants to determine the "improper" integral of a density ψ which has a singularity at a point $\mathfrak{o}$. One computes, in the way described, the integral of ψ over the punched surface and then lets the radius of the circular hole $\mathfrak{K}$, with center $\mathfrak{o}$, tend to zero.

The following lemma is of great consequence in the integration theory of densities.

Lemma L. If $d\eta = \operatorname{grad} \eta$ *is an exact differential and if* df *is a closed continuous differential which is also compact, then*

$$\int [d\eta, df] = 0. \tag{10.6}$$

We give the proof for closed surfaces, and add at the end the modifications which must be made if the assumption of compactness is transferred from the surface $\mathfrak{F}$ to the differential df.

The notations B_i, B_i^0, x_i, y_i, and μ_i will be used as above; we shall assume that the Dieudonné factors μ_i are continuously differentiable. Then the left-hand side of (10.6) is the sum of the integrals

$$\iint\limits_{B_i} \mu_i \left\{ \frac{\partial \eta}{\partial x_i} f_{y_i} - \frac{\partial \eta}{\partial y_i} f_{x_i} \right\} dx_i\, dy_i. \tag{10.7}$$

I claim that the individual integral (10.7) is equal to

$$-\iint\limits_{B_i} \eta \left\{ \frac{\partial \mu_i}{\partial x_i} f_{y_i} - \frac{\partial \mu_i}{\partial y_i} f_{x_i} \right\} dx_i\, dy_i. \tag{10.8}$$

Since μ_i vanishes outside B_i, the integral (10.8) is equal to the integral over all of $\mathfrak{F}$ of the density

$$-\eta\, [\operatorname{grad} \mu_i, df].$$

The sum of these last integrals is 0, for $\Sigma \mu_i = 1, \Sigma \operatorname{grad} \mu_i = 0$.

If one sets $\mu\eta = \lambda$, then to transform (10.7) into (10.8) one must show that

$$\iint_B \left\{\frac{\partial \lambda}{\partial x} f_y - \frac{\partial \lambda}{\partial y} f_x\right\} dx\, dy = 0 \tag{10.9}$$

for a continuously differentiable λ that vanishes outside of the block B, (10.4), and for a continuous closed $df = f_x\, dx + f_y\, dy$ in B. We choose a large natural number $N = 1/\varepsilon$ and divide the block in $4N^2$ small rectangles of dimensions a/N and b/N,

$$B_\varepsilon : x_0 \leq x \leq x_0 + a\varepsilon, y_0 \leq y \leq y_0 + b\varepsilon .$$

Since B is compact, N can certainly be chosen so large that the integral of the closed df around the perimeter of each small block B_ε vanishes. If one approximates λ by the first terms of its Taylor development about the vertex (x_0, y_0),

$$\lambda = \lambda_0 + \lambda_x^0 (x - x_0) + \lambda_y^0 (y - y_0) + \varepsilon\, o(\varepsilon),$$

then the integral of λdf around the perimeter of B_ε is approximated by

$$\lambda_0 \int df + \int \{\lambda_x^0 (x - x_0) + \lambda_y^0 (y - y_0)\} (f_x^0\, dx + f_y^0\, dy) \tag{10.10}$$

with an error $\varepsilon^2\, o(\varepsilon)$, and uniformly for all the $4N^2$ subrectangles B_ε. The first term drops out; in the second term f_x^0 and f_y^0 naturally denote the values of f_x and f_y at the vertex (x_0, y_0). So we get for (10.10)

$$\left\{\frac{\partial \lambda}{\partial x} f_y - \frac{\partial \lambda}{\partial y} f_x\right\}_{\substack{x = x_0 \\ y = y_0}} ab\, \varepsilon^2 .$$

Summing over all rectangles and letting $N \to \infty$ one obtains

$$\int \lambda\, df = \iint_B \left(\frac{\partial \lambda}{\partial x} f_y - \frac{\partial \lambda}{\partial y} f_x\right) dx\, dy, \tag{10.11}$$

where the left-hand integral is over the boundary of B. Since λ vanishes on the boundary of B, this gives the equation (10.9) which was to be proved. But the more general relation (10.11) (Green-Stokes formula) which does not assume that λ vanish on the boundary is important for us. An analogous equation is valid in an annulus R:

$$\int_R [\operatorname{grad} \lambda, df] = \int \lambda\, df . \tag{10.12}$$

On the right is the difference of the integrals over the two boundaries of the annulus.

If the hypothesis that $\mathfrak{F}$ is compact is dropped and replaced by the hypothesis that df is compact, let G be a compact set outside of which $df = 0$. One must go far enough in the sequence of Dieudonné factors μ so that $\mu_1 + \cdots + \mu_n = 1$ in all of G in order to carry over the proof unchanged.

Punch out from $\mathfrak{F}$ the disc $K_0 : r \leq a_0$ about the point $\mathfrak{o}$ with admissible local coordinates x, y. Let η be a function which is defined and continuously differentiable on the remainder $\mathfrak{F} - K_0$, and let df be a continuous closed differential on $\mathfrak{F} - K_0$. We also assume that df vanishes identically outside a compact part of $\mathfrak{F}$ (which contains K_0). We speak then of a closed *differential* on $\mathfrak{F} - K_0$ which is *compact relative to* $\mathfrak{F}$. Choose two z-discs K and K' with radii a and a' such that $a_0 < a < a'$ and a smoothing function λ for the annulus (a, a'). By means of $\mu = 1 - \lambda$, the integral of $[\operatorname{grad} \eta, df]$ over the punched surface $\mathfrak{F}' = \mathfrak{F} - K$ with boundary κ may be expressed as the sum of the integral of $[\operatorname{grad} \mu\eta, df]$ over the whole of $\mathfrak{F}$ and the integral of $[\operatorname{grad} \lambda\eta, df]$ over the annulus (a, a'). The second of these integrals is to be computed in polar coordinates. By Lemma L, whose proof does not become invalid under the situation described, the first integral vanishes. By Stokes' formula for the annulus, the second integral has the value $-\int_\kappa \eta \, df$. Thus in place of (10.6) we have the formula

$$\int_{\mathfrak{F}'} [\operatorname{grad} \eta, df] = -\int_\kappa \eta \, df. \tag{10.13}$$

If $\mathfrak{o}_1, \ldots, \mathfrak{o}_n$ are several points and if a circular hole with boundary κ_i is punched out about $\mathfrak{o}_i$, then the equation

$$\int_{\mathfrak{F}'} [\operatorname{grad} \eta, df] = -\sum_i \int_{\kappa_i} \eta \, df \tag{10.13}$$

holds under analogous conditions. (Here it is assumed that the holes are disjoint, and of course we speak of a "circle" κ_i relative to some admissible local coordinate system at $\mathfrak{o}_i$.)

A special case, obtained by setting $\eta = 1$, is the general *residue theorem*. This applies in the following situation. Let $\mathfrak{F}$ be a smooth oriented surface and let $\mathfrak{F}^\cdot$ be the surface obtained by deleting the n points $\mathfrak{o}_1, \ldots, \mathfrak{o}_n$ from $\mathfrak{F}$. Let a closed differential df be given on $\mathfrak{F}^\cdot$ and let df be compact relative to $\mathfrak{F}$. The conclusion of the theorem is: *the sum of the residues of df at the n "singularities"* $\mathfrak{o}_i$ *is equal to zero.*

§ 11. The intersection number

Put roughly, the *intersection number* $\operatorname{ch}(\alpha, \beta)$ of two closed paths α and β on a smooth oriented surface $\mathfrak{F}$ is the algebraic sum of the crossings of α over β; in this sum, a crossing from left to right contributes $+1$, a crossing from right to left contributes -1. The traffic rule on the "right of way" rests on the fact that a crossing of β by α from left to right is at the same time a crossing of α by β in the opposite sense. Thus the intersection number is skew symmetric in α and β:

$$\operatorname{ch}(\beta, \alpha) = -\operatorname{ch}(\alpha, \beta). \tag{11.1}$$

When one considers all the possibilities which the general concept of a continuous curve leaves open, it is clear that the definition in the above rough form is unusable. To arrive at a rigorous formulation we start with the following problem. For any two points $\mathfrak{a}$ and $\mathfrak{b}$ on $\mathfrak{F}$ which are "close together" and for a closed curve γ which does not pass through $\mathfrak{a}$ or $\mathfrak{b}$, to define the number which tells "how many times in all γ passes between $\mathfrak{a}$ and $\mathfrak{b}$ in the positive sense."

Let $\mathfrak{o}$ be a point of $\mathfrak{F}$, (x, y) admissible local coordinates at $\mathfrak{o}$, and $r < a_0$ a corresponding disc neighborhood $\mathfrak{E}$. It is no restriction to assume that $a_0 = 1$. Further let 1 and 2 be two points in $\mathfrak{E}$ and let $\phi_1(p)$ and $\phi_2(p)$ be the angles which the rays $\overrightarrow{1p}$ and $\overrightarrow{2p}$ form with the positive x-axis; here $p = (x, y)$ is any point of the xy-plane which is distinct from 1 and 2. The angles ϕ_1 and ϕ_2 are only determined mod 1; but we can follow their continuous variation along arbitrary curves γ which do not pass through 1 or 2. The difference $\psi = \phi_2 - \phi_1$ measures the angle through which one must rotate the direction $\overrightarrow{p1}$ in the positive sense about p to carry it into $\overrightarrow{p2}$. Since the azimuth ϕ of the ray $\overrightarrow{Op}$ satisfies the equation

$$2\pi d\phi = \frac{y\,dx - x\,dy}{x^2 + y^2},$$

$d\psi$ is a continuous closed differential in the unit disc $\mathfrak{E}$ with the points 1 and 2 deleted; also $d\psi$ has a residue $+1$ at the point 2 and a residue -1 at the point 1. It is exact modulo 1 and is infinite of the first order at the points 1 and 2.

We choose an $a < 1$ such that the concentric disc $\mathfrak{K}$, $r^2 = x^2 + y^2 < a^2$, contained in $\mathfrak{E}$, contains the points 1 and 2 in its interior. On the periphery of $\mathfrak{K}$, $\psi(p)$ is a uniform continuously differentiable function (which remains in absolute value $< \frac{1}{2}$); ψ may be extended to a function $\bar{\psi}$ with the same properties on the punched surface $\mathfrak{F} - \mathfrak{K}$. In particular, let this be done in such a way that $\bar{\psi}$ vanishes outside $\mathfrak{E}$. One proceeds most simply as follows. Enclose $\mathfrak{K}$ in a concentric disc $\bar{\mathfrak{K}}$ of radius $\bar{a} > a$ and < 1, and form the function $\bar{\psi} = \lambda\psi$ where λ is a (continuously differentiable) smoothing function for the annulus $(a, \bar{a})$ [where $\bar{\mathfrak{K}}$ sticks out beyond $\mathfrak{K}$]. The function $\bar{\psi}$ is defined outside $\mathfrak{K}$. The differential $d\psi_{12}$, which is equal to $d\psi$ in $\mathfrak{K}$ and equal to $d\bar{\psi}$ outside $\mathfrak{K}$, is a continuous closed differential on the surface with the points *1* and *2* deleted; it is exact mod 1. Thus one can form the integral $\int_\gamma d\psi_{12}$ along any curve γ on $\mathfrak{F}$ which does not pass through the points *1* or *2*. For a *closed* curve γ its value is $\equiv 0 \pmod 1$, that is, an *integer*. We are tempted to say that this number tells how many times the curve γ passes between *1* and *2*. But we must beware the fact that besides the points *1* and *2* all sorts of constructions go into the definition.

If γ stays outside $\mathfrak{E}$, then $\int_\gamma d\psi_{12} = 0$.

Let df be a continuous closed compact differential on $\mathfrak{F}$ and let α_{12} be any curve in $\mathfrak{E}$ joining the points *1* and *2*. Then I claim that the formula

$$\int_{\mathfrak{F}} [d\psi_{12}, df] = \int_{\alpha_{12}} df \tag{11.2}$$

is valid. Because of the singularities of $d\psi_{12}$ at *1* and *2*, the integral on the left-hand side is an improper integral. For the proof we can restrict ourselves to the interior of $\mathfrak{E}$, for $\psi_{12} = \bar{\psi}$ vanishes outside $\bar{\mathfrak{K}}$. Remove small discs κ_1^ε and κ_2^ε of radius ε about *1* and *2*, and let $\mathfrak{E}'$ denote the punched unit disc (that is, with the two small discs removed). On $\mathfrak{E}$, df is the differential of a uniform continuously differentiable function f, and the formula (10.13) yields

$$\int_{\mathfrak{E}'} [d\psi_{12}, df] = -\int_{\mathfrak{E}'} [df, d\psi_{12}] = \int_{\kappa_1^\varepsilon} f\, d\psi_{12} + \int_{\kappa_2^\varepsilon} f\, d\psi_{12}.$$

From passage to the limit $\varepsilon \to 0$ we obtain

$$\int_{\mathfrak{F}} [d\psi_{12}, df] = f(2) - f(1),$$

which is in fact (11.2).

The next step is to piece together an arbitrarily extended curve α with small arcs such as α_{12}. Let α be a curve from $\mathfrak{a}$ to $\mathfrak{b}$ on $\mathfrak{F}$. After choosing an admissible local coordinate system at each point p of α and an associated disc neighborhood $\mathfrak{E}(p)$, we obtain a standard subdivision of α into arcs $\alpha_{12} + \alpha_{23} + \cdots + \alpha_{n-1,n}$, each of which satisfies the hypotheses on α_{12}. Thus for each of these arcs there is an admissible coordinate system x, y and an associated disc $\mathfrak{E}$ containing the arc. Using these coordinates, we form the differentials $d\psi_{12}, d\psi_{23}, \ldots$ for each individual arc. It follows from (11.2) that the sum

$$d\psi^* = d\psi_{12} + d\psi_{23} + \cdots$$

satisfies the equation

$$\int_{\mathfrak{F}} [d\psi^*, df] = \int_\alpha df.$$

But it is disturbing that $d\psi^*$ has singularities not only at the endpoints of α

but also at the division points *2*, *3*, ... We shall eliminate these by a simple smoothing process.[24] The point $\mathfrak{o} = 2$ is covered by the coordinates x, y associated with α_{12} as well as by the coordinates x', y' for α_{23}. Let $\mathfrak{K}'$ and ϕ_2' have the same meaning for x', y' as $\mathfrak{K}$ and ϕ_2 for x, y. We choose a neighborhood of $\mathfrak{o}$ which is contained in both $\mathfrak{K}$ and $\mathfrak{K}'$ and which is a disc $\mathfrak{k}$ about $\mathfrak{o}$ in the local coordinates $x - x_0$, $y - y_0$ at $\mathfrak{o}$. In $\mathfrak{K}$, $d\psi_{12} = d\phi_2 - d\phi_1$ and in $\mathfrak{K}'$, $d\psi_{23} = d\phi_3' - d\phi_2'$. The problem is to smooth the differential $d\phi_2 - d\phi_2'$ at $\mathfrak{o}$. Because of the invariance of the order under topological maps, $\int_\gamma d\phi_2' = \int_\gamma d\phi_2$ for every closed curve γ in $\mathfrak{k}$ which does not pass through $\mathfrak{o}$.

Therefore $\phi_2 - \phi_2' = \Phi$ is a uniform continuously differentiable function in $\mathfrak{k}^\cdot$ (the disc $\mathfrak{k}$ with $\mathfrak{o}$ removed). By means of a smoothing function λ which vanishes outside $\mathfrak{k}$ and is identically 1 in a smaller concentric disc $\mathfrak{k}'$, we obtain from Φ a function $\omega = \lambda\Phi$. The smoothed differential $d\Phi - d\omega$ is then regular at $\mathfrak{o}$, for it vanishes in a whole neighborhood of $\mathfrak{o}$. But by subtracting $d\omega$, the value of the integral $\int_\beta (d\psi_{12} + d\psi_{23})$ has not been changed for any closed curve β which does not pass through *1*, *2*, or *3*. Also the value of the surface integral of $[d\psi_{12} + d\psi_{23}, df]$ is unchanged, since $\int [d\omega, df] = 0$. For if we operate in the disc $\mathfrak{k}$ and use the polar coordinates associated with $x - x_0$, $y - y_0$, then the value of that integral over an annulus, bounded by $\mathfrak{k}$ and a smaller concentric circle $\mathfrak{k}_\varepsilon$ of radius ε, is $-\int_{\mathfrak{k}_\varepsilon} \omega\, df$ by Stokes' formula (10.12) for an annulus. It is easy to see that ω remains bounded in a neighborhood of $\mathfrak{o}$; but it suffices to observe that $d\omega$ is at most of first order at $\mathfrak{o}$ and hence that at worst ω itself becomes logarithmically infinite. Then it follows that the integral $\int_{\mathfrak{k}_\varepsilon} \omega\, df$ tends to zero at least as rapidly as $\varepsilon \log(1/\varepsilon)$, as $\varepsilon \to 0$.

After this smoothing we obtain $d\psi$ in place of $d\psi^*$, and the two differ by an exact differential $d\omega$. More exactly, ω is a function which is continuously differentiable except at a finite number of singular points (here the points *2*, *3*, ... , n − *1*) and whose differential $d\omega$ has an infinity of at most first order at each point. Such a function will be called an ω-function. This is enough to draw the conclusion $\int [d\omega, df] = 0$ for every closed compact df. In the case of a *closed* curve α, the smoothing is also to be carried out at the initial = terminal point, *1* = n. Finally we allow the addition to $d\psi$ of the differential of an arbitrary ω-function with singularities at the endpoints $\mathfrak{a}$ and $\mathfrak{b}$ of the curve α (for the case of a closed curve: the differential of an

---

[24]) On a Riemann surface, where we may always take $z = x + iy$ to be a local parameter, these singularities are certainly not present. In that case the smoothing process is superfluous.

arbitrary ω-function without singularities). Denote the resulting differential by ds. Then

$$\int [ds, df] = \int_\alpha df. \tag{11.3}$$

To simplify the terminology we shall understand by (α) a curve α together with the following constructions: a subdivision of α into arcs, each of which is contained in a unit disc $\mathfrak{E}$ relative to some definite local coordinates x, y. Then $ds = ds(\alpha)$ is uniquely determined by (α) in the following sense: any two differentials $ds(\alpha)$ and $ds^*(\alpha)$ corresponding to (α) satisfy the formula

$$ds(\alpha) \sim ds^*(\alpha) \qquad (\mathfrak{a}, \mathfrak{b}) \tag{11.4}$$

of "modified cohomology." It means that the difference $ds(\alpha) - ds^*(\alpha)$ is the differential of an ω-function whose singularities are at most the endpoints $\mathfrak{a}$ and $\mathfrak{b}$ of α. [A closed curve is without endpoints, and the qualification $(\mathfrak{a}, \mathfrak{b})$ behind formula (11.4) disappears.] If α is a curve from $\mathfrak{a}$ to $\mathfrak{b}$ and β is a curve from $\mathfrak{b}$ to $\mathfrak{c}$, then one can form not only $\alpha + \beta$ in the obvious way from α and β, but also $(\alpha + \beta)$ from (α) and (β). Obviously

$$ds(\alpha + \beta) \sim ds(\alpha) + ds(\beta) \qquad (\mathfrak{a}, \mathfrak{b}, \mathfrak{c})$$

holds.

Our undertaking, to arrive at an invariant definition of the intersection number, would be hopeless if $ds(\alpha)$ were not also independent, in the sense of cohomology, of the constructions associated with α. Fortunately the following theorem is valid.

If α and α' are two paths joining the same pair of points $\mathfrak{a}$ and $\mathfrak{b}$ and if the closed path $\gamma = \alpha - \alpha'$ is weakly homologous to zero – for which we write simply $\alpha \sim \alpha'$ –, then

$$ds(\alpha) \sim ds(\alpha') \qquad (\mathfrak{a}, \mathfrak{b}) \tag{11.5}$$

always holds.

In particular this is the case if α' and α coincide. *Thus for every closed path β which does not pass through $\mathfrak{a}$ or $\mathfrak{b}$, the integer*

$$\int_\beta ds(\alpha) = \mathrm{ch}(\beta, \alpha) \tag{11.6}$$

is uniquely determined by α and β in such a fashion that α may be replaced by any $\alpha' \sim \alpha$ without changing this integer. Then it is reasonable to call (11.6) the *intersection number* of the closed curve β with the curve α.

To prove our theorem we form $(\alpha) - (\alpha') = (\gamma)$. This (γ) belongs to the closed cruve γ which, by assumption, is weakly homologous to zero. Let β be a second closed curve, which by the related constructions gives a (β) and an associated differential $ds(\beta)$ that is everywhere continuous and closed, and is also compact. We form the density integral $\int[ds(\gamma), ds(\beta)]$. By our fundamental formula (11.3) this integral has the value $\int_\gamma ds(\beta)$ and hence is zero, since $\gamma = \alpha - \alpha' \sim 0$. On the other hand, by interchanging β and γ we obtain for the same integral the value

$$-\int[ds(\beta), ds(\gamma)] = -\int_\beta ds(\gamma).$$

The vanishing of the integral on the right-hand side for every closed path β amounts to $ds(\gamma) \sim 0$. Thus (11.5) follows.

From the observation that $\int_\gamma d\psi_{12} = 0$ in case the closed curve γ lies outside $\mathfrak{E}$ (p. 81), and by partitioning α into sufficiently fine arcs $\alpha_{12}, \ldots$ we obtain the following result. *The intersection number* $\mathrm{ch}(\beta, \alpha)$ *is always zero if the closed curve* β *does not meet the curve* α.

Now we can write ds_α in place of $ds(\alpha)$, indicating thereby that ds_α is determined uniquely, in the sense of the modified cohomology (11.4), by the path α. In this sense ds_α remains unchanged even if α is replaced by a path $\alpha' \sim \alpha$.

The intersection number (11.6) of two *closed* curves α and β is obviously unchanged if β is replaced by a path weakly homologous to β. Also, from the result obtained, it remains unaltered under a similar replacement of α. From the equation

$$\int_\beta ds_\alpha = \int_{\mathfrak{F}}[ds_\beta, ds_\alpha] = -\int_{\mathfrak{F}}[ds_\alpha, ds_\beta] = -\int_\alpha ds_\beta,$$

which has already been used, comes the *law of skew symmetry* (11.1). *The goal set at the beginning of this paragraph, a rigorous definition of the intersection number of closed curves, is thus attained.*

Before we turn to closed curves finally, we shall generalize our formulas to include the case where *not only* α *but also* β *is open.* Let $\mathfrak{b}$ and $\mathfrak{b}'$ be two points on $\mathfrak{F}$ and let df be a continuous closed differential, on the surface with the points $\mathfrak{b}$ and $\mathfrak{b}'$ deleted, which is compact relative to $\mathfrak{F}$. Let the residues of df at $\mathfrak{b}$ and $\mathfrak{b}'$ be $-A$ and $+A$, respectively. Let $\alpha = \mathfrak{a}\mathfrak{a}'$ be a curve which does not pass through $\mathfrak{b}$ or $\mathfrak{b}'$. We can choose the partition of α into arcs $\alpha_{12}, \alpha_{23}, \ldots$ such that the points $\mathfrak{b}$ and $\mathfrak{b}'$ lie outside all of the "discs" $\mathfrak{E}$ enclosing the arcs of α. Then we can use the formula (11.2) unchanged for the individual arcs α_{12}, for $d\psi_{12}$ vanishes outside $\bar{\mathfrak{K}}$ and everything takes place in the unit disc $\mathfrak{E}$, in which df has no singularity. Let $\mathfrak{b}$ and $\mathfrak{b}'$ be joined by a curve β which does not meet the points *1* and *2*. One can also find a curve β' which joins $\mathfrak{b}$ and $\mathfrak{b}'$ and does not penetrate the interior of $\bar{\mathfrak{K}}$. Namely, note the first and last intersections of β with the periphery $\bar{\kappa}$ of $\bar{\mathfrak{K}}$. If those points are $\mathfrak{c}$ and $\mathfrak{c}'$, then replace that portion of β from $\mathfrak{c}$ to $\mathfrak{c}'$ by the arc of $\bar{\kappa}$ from $\mathfrak{c}$ to $\mathfrak{c}'$. Then $\int_{\beta'} d\psi_{12} = 0$, for ψ_{12} vanishes outside $\bar{\mathfrak{K}}$. Thus $\int_\beta d\psi_{12}$ is equal to the integral of $d\psi_{12}$ over the *closed* path $\beta - \beta'$ and hence is an integer. Summation and subsequent smoothing yields the relations

$$\int [d\psi, df] = \int_\alpha df, \qquad \int_\beta d\psi \equiv 0 \qquad (\mathrm{mod}\, 1).$$

Now we may take β to be an arbitrary path from $\mathfrak{b}$ to $\mathfrak{b}'$ which does not pass through the endpoints $\mathfrak{a}$ and $\mathfrak{a}'$ of α; it is no longer necessary that β avoid the division points *2*, *3*, ... A ds_α arises from $d\psi$ by the addition of the differential $d\omega$ of an ω-function which has singularities only at the points $\mathfrak{a}$ and $\mathfrak{a}'$. On the one hand,

$$\int_\beta ds_\alpha \equiv \omega(\mathfrak{b}') - \omega(\mathfrak{b}) \qquad (\mathrm{mod}\, 1).$$

On the other hand, the formula (10.13) applied to $\eta = \omega$ and two circles κ shrinking down to $\mathfrak{b}$ and $\mathfrak{b}'$ yields

$$\int [d\omega, df] = A\{\omega(\mathfrak{b}) - \omega(\mathfrak{b}')\}. \tag{11.7}$$

Thus the equation

$$\int [ds_\alpha, df] + A \int_\beta ds_\alpha \equiv \int_\alpha df \qquad (\mathrm{mod}\, A) \tag{11.8}$$

follows. From these considerations themselves it follows that if df and β are given, then the left-hand side of (11.8) remains unchanged if the construction of ds_α is altered, so long as α remains fixed. For the difference of two such ds_α is the differential of an ω-function which has singularities only at $\mathfrak{a}$ and $\mathfrak{a}'$!

If we take now for df the differential ds_β associated with a curve β from $\mathfrak{b}$ to $\mathfrak{b}'$, then we see that

$$\mathrm{ch}(\beta, \alpha) = \int [ds_\alpha, ds_\beta] - \{\int_\alpha ds_\beta - \int_\beta ds_\alpha\} \tag{11.9}$$

is an integer which for a fixed choice of ds_β is uniquely determined by α. Because of the skew symmetry of this expression in α and β, it depends only on β and not on the special construction of ds_β. Therefore *the integer* (11.9), *uniquely determined by α and β, may be called the intersection number of the two open paths α and β. It is* 0 if α and β are disjoint. If β is closed, and therefore ds_β has no singularity, then, by virtue of

$$\int [ds_\alpha, ds_\beta] = \int_\alpha ds_\beta,$$

(11.9) gives the former value $\int_\beta ds_\alpha$.

If, instead of df, one applies the formula (11.8) to the differential $df - A\,ds_\beta$, which has the residue 0 at $\mathfrak{b}$ and $\mathfrak{b}'$, then

$$\int [ds_\alpha, df] + A\{\int_\beta ds_\alpha - \operatorname{ch}(\beta, \alpha)\} = \int_\alpha df, \tag{11.10}$$

from which the *congruence* (11.8) mod A is turned into an *equation*. The difference $\int_\beta ds_\alpha - \operatorname{ch}(\beta, \alpha)$ remains unaltered if β is replaced by a path β^* joining the same two points $\mathfrak{b}$ and $\mathfrak{b}'$, and therefore will be written in abbreviated form as $[s_\alpha]_{\mathfrak{b}}^{\mathfrak{b}'}$. Naturally, this must be so, for the expression $\int [ds_\alpha, df] - \int_\alpha df$ contains df but not a path β joining its two singularities $\mathfrak{b}$ and $\mathfrak{b}'$.

With an eye to the function theoretic applications, we supplement equation (11.10) with the following. Let u_α be a continuously differentiable function everywhere except at the endpoints $\mathfrak{a}$ and $\mathfrak{a}'$ of α, and let du_α have infinities of at most the first order at $\mathfrak{a}$ and $\mathfrak{a}'$. Then the equation

$$\int [du_\alpha, df] + A[u_\alpha]_{\mathfrak{b}}^{\mathfrak{b}'} = 0 \tag{11.11}$$

is valid, where df is a differential satisfying the conditions specified for (11.10). Here $[u_\alpha]_{\mathfrak{b}}^{\mathfrak{b}'}$ denotes the difference $u_\alpha(\mathfrak{b}') - u_\alpha(\mathfrak{b})$. This is nothing but the equation (11.7) in different notation. If df possesses not just two but several singularities $\mathfrak{b}_1, \ldots, \mathfrak{b}_n$ with the residues $A_1, \ldots, A_n$, then choose a point $\mathfrak{o}$ not on α. Instead of (11.10) and (11.11) one obtains the more general equation

$$\int_{\mathfrak{F}} [du_\alpha, df] + \sum_{i=1}^{n} A_i [u_\alpha]_{\mathfrak{o}}^{\mathfrak{b}_i} = 0,$$

$$\int_{\mathfrak{F}} [ds_\alpha, df] + \sum_{i=1}^{n} A_i [s_\alpha]_{\mathfrak{o}}^{\mathfrak{b}_i} = \int_\alpha df.$$

The remainder of this paragraph is concerned only with *closed* curves. Let α be one such. It is not only true that the weak homology $\alpha \sim 0$ implies $\operatorname{ch}(\alpha, \beta) = 0$ for every closed curve β, but the important converse is valid.

Theorem T. *If the closed path α satisfies the equation* $\operatorname{ch}(\alpha, \beta) = 0$ *for every closed path β, then α is weakly homologous to zero.*

Since that equation can be written in the form $\int_\beta ds_\alpha = 0$, the assertion may be expressed as follows: the weak homology $\alpha \sim 0$ and the cohomology

$ds_\alpha \sim 0$ are equivalent. It is apposite to formulate this theorem more generally as follows.

If $\alpha_1, \dots, \alpha_l$ *are closed paths and if* $c_1, \dots, c_l$ *are real constants, then the weak homology*

$$c_1\alpha_1 + \cdots + c_l\alpha_l \sim 0 \tag{11.12}$$

and the cohomology

$$ds = c_1\, ds_{\alpha_1} + \cdots + c_l\, ds_{\alpha_l} \sim 0$$

are equivalent.

Proof. The relation (11.12) is by definition equivalent to the statement that the equation

$$c_1 \int_{\alpha_1} df + \cdots + c_l \int_{\alpha_l} df = 0$$

holds for every closed compact differential df. But this may be written in the form

$$c_1 \int [ds_{\alpha_1}, df] + \cdots + c_l \int [ds_{\alpha_l}, df] = 0$$

or

$$\int [ds, df] = 0. \tag{11.13}$$

So we must show that $ds \sim 0$ implies (11.13) for every closed compact differential df; and conversely, that if (11.13) holds for every such differential, then $ds \sim 0$. The first part is identical with Lemma L. One gets the converse by picking df to be ds_β and then concluding, by virtue of $\int [ds, ds_\beta] = 0$, that

$$\int_\beta ds = \int [ds_\beta, ds] = -\int [ds, ds_\beta] = 0$$

for each closed path β, and therefore $ds \sim 0$.

If we regard closed *streams*

$$\sigma \sim c_1\alpha_1 + \cdots + c_l\alpha_l$$

as equal if they are weakly homologous to each other, then according to our theorem there corresponds to each such stream σ a unique (in the sense of cohomology) differential

$$ds_\sigma = c_1\, ds_{\alpha_1} + \cdots + c_l\, ds_{\alpha_l}.$$

Now we shall assume that *the weak degree of* $\mathfrak{F}$, the maximal number of closed paths which are linearly independent in the sense of weak homology, *is finite*; we shall now denote this number by h. If $\gamma_1, \dots, \gamma_h$ are any h closed paths which are linearly independent in the specified sense, then every closed path α is weakly homologous to a linear combination $x_1\gamma_1 + \cdots + x_h\gamma_h$

with real coefficients x_i. Thus if df is any closed compact differential on $\mathfrak{F}$, then the equation

$$\int_\alpha df = \sum_{j=1}^h x_j \int_{\gamma_j} df$$

holds. Thus $df \sim 0$ if the h equations

$$\int_{\gamma_1} df = 0, \cdots, \int_{\gamma_h} df = 0$$

are satisfied.

By the general theorem we have just proved, the h differentials $ds_i = ds_{\gamma_i}$ are linearly independent in the sense of cohomology. That is, if the equation

$$\sum_{j=1}^h c_j \int_\alpha ds_j = 0$$

holds for all closed paths α, then we must have $c_1 = \cdots = c_h = 0$. Thus the h linear equations

$$\sum_{j=1}^h c_j s_{ij} = 0 \qquad (i = 1, \dots, h)$$

for h unknows c_j, with integral coefficients $s_{ij} = \int_{\gamma_i} ds_j$ have only the trivial solution $c_1 = \cdots = c_h = 0$. In other words, the determinant Δ of the s_{ij} is not zero; or, the skew-symmetric bilinear form (*characteristic form*)

$$\sum_{i,j=1}^h s_{ij} x_i y_j, \tag{11.14}$$

which gives the value of

$$\mathrm{ch}(\alpha, \beta) = \int [ds_\alpha, ds_\beta]$$

for the two streams

$$\alpha \sim x_1\gamma_1 + \cdots + x_h\gamma_h, \qquad \beta \sim y_1\gamma_1 + \cdots + y_h\gamma_h,$$

is nondegenerate.

Since the equations

$$\sum_j s_{ij} x_j = a_i \qquad (i,j = 1, \dots, h)$$

always have a unique solution, no matter what real numbers a_i are given, the periods $\int_{\gamma_i} df$ of a closed compact differential $df = x_1\, ds_1 + \cdots + x_h\, ds_h$ may be specified arbitrarily; then df is uniquely determined in the sense of cohomology.

For a closed *path* α,

$$\alpha \sim x_1\gamma_1 + \cdots + x_h\gamma_h, \tag{11.15}$$

the equations

$$\mathrm{ch}(\gamma_i, \alpha) = \sum_{j=1}^h s_{ij} x_j$$

hold and may be used to compute the x_j. Since the left-hand members and the coefficients are all integers, it follows that *the x_i are necessarily rational numbers with denominator* Δ.

But a skew-symmetric bilinear form can be nondegenerate only if the dimension number h is even: *if the weak degree h is finite, it is even.* For the transposed matrix $s'_{ij} = s_{ji}$ has the same determinant Δ as s_{ij}. If $s'_{ij} = -s_{ij}$, then also $\Delta = (-1)^h\Delta$, which is compatible with $\Delta \neq 0$ only if h is even. For closed oriented smooth surfaces we set $h = 2p$ and call p the *genus* of the surface.

Relative to the basis $\gamma_1, \ldots, \gamma_h$, every closed curve α is, by (11.15), characterized by the vector $x = (x_1, \ldots, x_h)$ in an h-dimensional vector space. These vectors form a discontinuous *lattice* G, i.e., a group under addition. In other words: the zero vector belongs to the lattice; if x belongs to G, then $-x$ belongs to G; if two vectors x and x' belong to G, so does their sum $x + x'$. The coordinates x_i of a lattice vector are rational numbers with the universal denominator Δ. On the other hand, the unit lattice consisting of all vectors with integral coordinates x_i is contained in G. By a construction which Minkowski called the "adaptation of a number lattice relative to a sublattice,"[25] one can choose the basis $\gamma_1, \ldots \gamma_h$ such that for every closed path α the coefficients x_i in the weak homology (11.15) are *integers*; that is, so that G coincides with the unit lattice. We speak then of an *integral basis* for the closed paths. The passage from one such basis to another is accomplished by a unimodular transformation; that is, a linear transformation A with integral coefficients and determinant ± 1. [Since A^{-1} also has integral coefficients, the last follows from the equation $\det(A) \cdot \det(A^{-1}) = 1$.]

The construction of an integral basis goes as follows. Among the finitely many vectors of the lattice of the form

$$(r_1, 0, \ldots, 0) \qquad (0 < r_1 \leq 1)$$

let

$$\gamma_1 \sim (r_1^{(1)}, 0, \ldots, 0)$$

be that one for which r_1 has the smallest value. Among the finitely many lattice vectors of the form

$$(r_1, r_2, 0, \ldots, 0) \qquad (0 \leq r_1 < r_1^{(1)}, 0 < r_2 \leq 1)$$

let

$$\gamma_2 \sim (r_1^{(2)}, r_2^{(2)}, 0, \ldots, 0)$$

be that one for which r_2 is as small as possible. Among all lattice vectors

$$(r_1, r_2, r_3, 0, \ldots, 0) \qquad (0 \leq r_1 < r_1^{(1)}, 0 \leq r_2 < r_2^{(2)}, 0 < r_3 \leq 1)$$

---

[25]) See H. Minkowski, *Diophantische Approximationen*, Leipzig 1907, pp. 90–95.

let

$$\gamma_3 \sim (r_1^{(3)}, r_2^{(3)}, r_3^{(3)}, 0, \leq, 0)$$

be that one for which r_3 is as small as possible. And so on. Then the vectors $\gamma_1, \dots, \gamma_h$ form an integral basis for the lattice. For if $\alpha \sim (r_1, \dots, r_h)$ is an arbitrary vector of the lattice, then one can determine the integers n_h, $n_{h-1}, \dots, n_1$ successively so that for

$$\alpha - (n_h \gamma_h + n_{h-1} \gamma_{h-1} + \cdots + n_1 \gamma_1) \sim (\bar{r}_1, \dots, \bar{r}_{h-1}, \bar{r}_h)$$

the inequalities

$$0 \leq \bar{r}_i < r_i^{(i)} \qquad (i = h, h-1, \dots, 1)$$

hold. But then one concludes from the definition of $\gamma_h, \gamma_{h-1}, \dots, \gamma_1$ that

$$\bar{r}_h = 0, \bar{r}_{h-1} = 0, \dots, \bar{r}_1 = 0,$$

and the end result establishes our claim:

$$\alpha \sim n_h \gamma_h + \cdots + n_2 \gamma_2 + n_1 \gamma_1 .$$

If $S_1, \dots, S_h$ are those cover transformations of the class surface $\hat{\mathfrak{F}}$ of $\mathfrak{F}$ which correspond to the paths $\gamma_1, \dots, \gamma_h$ of the integral basis, then the Abelian group of cover transformations of $\hat{\mathfrak{F}}$ consists of all transformations of the form $S_1^{n_1} \cdots S_h^{n_h}$, where the n_i are arbitrary integers. Thus this group is a free Abelian group with h generators.

If we compute the characteristic form (11.14) relative to an integral basis $\gamma_1, \dots, \gamma_h$, then the form is determined to within a unimodular simultaneous transformation of the variables x_i and y_i. If we call two forms equivalent and put them in the same class in case they are related by such a transformation, then the class to which the characteristic form belongs is determined by the surface. But, at least for closed surfaces, the characteristic form can always be put in the normal form

$$(x_1 y_2 - x_2 y_1) + \cdots + (x_{h-1} y_h - x_h y_{h-1}) \tag{11.16}$$

by an appropriate choice of the integral basis. Thus the form leads to no topological invariants beyond $h = 2p$. An integral basis relative to which the characteristic form is given by (11.16) is called a *canonical basis*. The proof of this theorem consists of a topological part and an arithmetical part. The topological part says that the determinant Δ of the characteristic form (which, as for every skew-symmetric form, must be positive) is equal to 1. The easily proved arithmetic part says that a skew-symmetric form with integral coefficients and determinant 1 may be transformed unimodularly into (11.16). The topological part may be expressed in two ways.

(1) If $n_1, \dots, n_h$ are arbitrary integers, then there exists a closed path α such that
$$\mathrm{ch}(\alpha, \gamma_1) = n_1, \dots, \mathrm{ch}(\alpha, \gamma_h) = n_h.$$

(2) To each primitive closed path α there corresponds a closed path β such that $\mathrm{ch}(\alpha, \beta) = 1$. Here α is called *primitive* if there does not exist a closed path α' such that α is homologous to a multiple $n\alpha'$ of α' (with $n = 2$, or 3, or 4, or ...).

We shall not prove these facts for closed surfaces at this point; nor shall we prove the other theorem, that *a completely planar closed surface is necessarily simply connected.* Both theorems are relatively easy results in combinatorial topology; but this is not on our route, since we would have to introduce the assumption that the surface is triangulable. We prefer to prove these results under the assumption that the surface is a *Riemann* surface. In this form they will automatically fall into our laps as corollaries of the theory of Riemann surfaces.

Riemann based the topological investigation of a closed oriented surface $\mathfrak{F}$ of genus p on the construction of p canonical pairs of cuts [Rückkehrschnittpaaren] $\pi_1, \pi'_1; \dots; \pi_p, \pi'_p$. These are simple closed curves; two curves belonging to different pairs do not intersect; π_i and π'_i intersect in exactly one point $\mathfrak{q}_i$, at which π'_i crosses π_i from left to right. Thus we have a canonical basis with special properties. Furthermore Riemann joined a point $\mathfrak{o}$ of the surface, not on any of the canonical pairs, to each of the points $\mathfrak{q}_i$ by simple disjoint paths σ_i not intersecting the canonical pairs, and showed that $\mathfrak{F}$ was turned into a simply connected domain by cutting it along the "reins" σ_i and the canonical pairs π_i, π'_i. To make this procedure rigorous one must start with a triangulation of the surface and specialize all the curves to polygons on the triangulated surface. I proceeded in that way in the first edition of this book. By means of this dissection one can show that *the class surface is the strongest of all normal perfect covering surfaces for which the group of cover transformations is Abelian.* For a curve on any such covering surface whose trace on $\mathfrak{F}$ is closed and homologous to zero is necessarily closed.[26]

As Möbius and C. Jordan have proved,[27] the genus is the only topological invariant of closed oriented smooth surfaces. Any two such surfaces of the

---

[26]) See page 176 of the second edition (1923).

[27]) Möbius, *Theorie der elementaren Verwandtschaft* (1863), *Werke* II, 435–471; C. Jordan, *Journal de Mathématiques,* ser. 2, **11** (1886), 105. Also see W. Dyck, *Math. Ann.,* **32** (1888) 457.

same genus are equivalent, not merely topologically, but as smooth surfaces: one may be mapped onto the other by a one-to-one transformation which, together with its inverse, is continuously differentiable. It is easy to give closed oriented surfaces in three-dimensional Euclidean space which have a specified genus p. For example, a sphere with p handles has genus p. The incisive significance of the genus of a closed Riemann surface for the theory of functions on this surface will become sufficiently clear from the function theoretic theorems of the next chapter.

II. FUNCTIONS ON RIEMANN SURFACES

§ 12. The Dirichlet integral and harmonic differentials

With a differential $df = f_x\,dx + f_y\,dy$, continuous at the point $\mathfrak{p}_0$ of the Riemann surface $\mathfrak{F}$, there is associated the conjugate differential

$$df' = -f_y\,dx + f_x\,dy; \qquad f'_x = -f_y, f'_y = f_x.$$

This association is invariant under change of the local parameter $z = x + iy$. If both df and df' are exact at the point $\mathfrak{p}_0$, then df is said to be *harmonic* there; $df + i\,df'$ is then regular analytic at $\mathfrak{p}_0$. Conversely, the real and imaginary parts of a differential which is regular analytic at $\mathfrak{p}_0$ are harmonic differentials. From the density $[df, dg] = f_x g_y - f_y g_x$ formed from two differentials one obtains, by replacing dg by its conjugate dg', the symmetric expression

$$[df, dg'] = (df, dg) = f_x g_x + f_y g_y = (dg, df).$$

Thus this quantity behaves as a density relative to transformations of the local parameter, and one can form the invariant integral $\int(df, dg)$ on a Riemann surface. For $dg = df$,

$$(df, df) = (df)^2 = f_x^2 + f_y^2$$

is nonnegative, and hence the integral $\int(df)^2$ exists on noncompact as well as on compact surfaces $\mathfrak{F}$, provided that the value ∞ is permitted in the first case. From the inequality

$$(a + b)^2 \leq 2(a^2 + b^2),$$

which is valid for real numbers a and b, it follows that if $\int(df)^2$ and $\int(dg)^2$ are finite, then so is $\int(df + dg)^2$. Then from

$$2(df, dg) = (df + dg)^2 - (df)^2 - (dg)^2$$

it follows that $\int(df, dg)$ exists and is finite. Because of the positive character of the quadratic form

$$\int(\xi\, df + \eta\, dg)^2 = \xi^2 \int(df)^2 + 2\xi\eta \int(df, dg) + \eta^2 \int(dg)^2 \geq 0$$

in the real variables ξ and η, the Cauchy-Schwarz inequality

$$\{\int(df, dg)\}^2 \leq \int(df)^2 \cdot \int(dg)^2 \tag{12.1}$$

is valid.

If df is exact, the gradient grad u of a continuously differentiable function u of points on the Riemann surface, then one can form

$$D(u) = \int_{\mathfrak{F}} (\operatorname{grad} u)^2 . \tag{12.2}$$

For two such functions u and v we have the corresponding symmetric bilinear functional

$$D(u, v) = \int(\operatorname{grad} u, \operatorname{grad} v) .$$

The integral $D(u)$, the "field energy," was used since 1847 by *W. Thomsen* (Lord *Kelvin*), and then by Lejeune *Dirichlet* in his lectures, in many problems in mathematical physics. Here the controlling natural laws appeared in the following form: for the actual field the energy is a minimum. Riemann constructed the functions and the differentials on a closed Riemann surface with the aid of this minimum principle, which was familiar to him from Dirichlet's lectures. In accord with the precedent set by Riemann, it has become customary to join the integral (12.2) with Dirichlet's name and to tag existence proofs based on finding that function, among the admitted functions, which minimizes the integral with the label *Dirichlet principle.*

Then Weierstrass' criticism[1] removed the foundation of such proofs; Weierstrass showed by examples that the existence of a minimizing function is by no means certain. Hence one looked around for other methods. *H. A. Schwarz* and *C. Neumann* invented the "alternating process" and in their hands it served brilliantly as a foundation for Riemannian function theory.[2]

---

[1]) See Weierstrass, *Werke,* **2,** 49–54, *Über das sog. Dirichletsche Prinzip* (1870).

[2]) H. A. Schwarz: *Über einen Grenzübergang durch alternierendes Verfahren,* Gesammelte Abh., Vol. II, 133–143; *Über die Integration der partiellen Differentialgleichung* $\Delta u = 0$ *unter vorgeschriebenen Grenz- und Unstetigkeits-Bedingungen,* ibid., 144–171; *Zur Integration der partiellen Differentialgleichung* $\Delta u = 0$, ibid., 175–210. C. Neumann: Ber. Sächs. Akad. d. Wiss. Leipzig, 1870; *Vorlesungen über Riemann's Theorie der Abelschen Integrale,* 2nd edition, Leipzig 1884, 388–471. Among the more recent literature I mention in particular the following two works of C. Carathéodory: *Elementarer Beweis für den Fundamentalsatz der konformen Abbildung,* in the *Festschrift zum 50jährigen Doktorjubiläum von H. A. Schwarz,* Berlin 1914, 19–41; *Conformal Representation,* Cambridge Tracts in Math., **28,** 1932.

But about the turn of the century David *Hilbert* furnished a rigorous proof of the Dirichlet principle in a form sufficiently general for Riemann's purposes.[3] Since that time the Dirichlet principle has again been one of the most powerful tools of analysis. Here we shall follow Hilbert's method with certain simplifications which the intervening period has brought about.[4]

If one wishes to minimize a functional, such as $D(u)$, of the argument u, then naturally one must circumscribe the domain of the competing quantities u. In the case of the Dirichlet integral let it be assumed that if u is a competing function, then so is every function which differs from u by an everywhere continuously differentiable function w which vanishes outside a neighborhood of a point $\mathfrak{p}_0$. Let this neighborhood be mapped conformally onto the interior E of a circle in the complex $z = x + iy$ plane. That a minimizing function, if it exists at all, must be a *harmonic function* is customarily made plausible as follows. The equation

$$D(u + \varepsilon w) = D(u) + 2\varepsilon D(u, w) + \varepsilon^2 D(w)$$

holds for arbitrary real ε. Since this quantity must be at least $D(u)$ for arbitrary positive or negative ε, the "*first variation*" $D(u, w)$ must vanish. By means of Green's formula this integral changes into

$$-\iint_E w \, \Delta u \, dx \, dy, \qquad \text{where} \qquad \Delta u = \frac{\partial^2 u}{\partial x^2} + \frac{\partial^2 u}{\partial y^2}.$$

---

[3]) See the two papers, *Über das Dirichletsche Prinzip*, in Hilbert's Gesammelte Abh., **3**, Berlin 1935, 10–14 and 15–37; and also his paper, *Zur Theorie der konformen Abbildung*, ibid., 73–80. The second was originally published in the *Festschrift zur Feier des 150jährigen Bestehens der Gesellschaft der Wissenschaften zu Göttingen*, 1901, and was reprinted in Math. Ann., **59** (1904). The other two originated in 1905 and 1909.

[4]) I cite here particularly the following early works which appeared before the first edition of this book: B. Levi, *Sul principio di Dirichlet*, Rend. Circ. Mat. Palermo, **22** (1906) 293–360; G. Fubini, *Il principio di minimo e i teoremi di esistenza per i problemi al contorno relativi alle equazione alle derivate parziali di ordini pari*, ibid., **23** (1907) 58–84; H. Lebesgue, *Sur le problème de Dirichlet*, ibid., **24** (1907) 371–402; S. Zaremba, *Sur le principe du minimum*, Bull. de l'Ac. des sciences de Cracovie, July 1909, 206ff; R. Courant, *Über die Methode des Dirichletschen Prinzips*, Math. Ann., **72** (1912) 517–550. What I give here, essentially unchanged, is the proof that I developed in 1913 on the basis of the work of Hilbert, B. Levi, and Zaremba; the attack is simpler than that of Riemann and Hilbert. The proof incorporates the modifications which are the natural consequence of Dieudonné's improved method of surface integration. Since that time a rich literature on the subject has accumulated. Through Courant, among others, the "direct methods of the calculus of variations" has come to be a familiar term in analysis. For a comprehensive presentation we refer to Courant's book on the Dirichlet principle, already mentioned in the preface.

From the fact that this must vanish for every function w of the type described, it follows that the equation $\Delta u = 0$ is valid in a neighborhood of $\mathfrak{p}_0$. To be sure, we have made use of the second derivatives of u, although it is not suitable for our purpose to include their existence among the conditions for competing functions. One will attempt to avoid this assumption by a small change in the argument, whose conclusion will be, not $\Delta u = 0$, but the existence of a conjugate function u' which is connected with u by the relations

$$\frac{\partial u}{\partial x} = \frac{\partial u'}{\partial y}, \qquad \frac{\partial u}{\partial y} = -\frac{\partial u'}{\partial x};$$

$u + iu'$ is then an analytic function of z. We shall now give a rigorous formulation of this heuristic argument which has served as an introduction.

That a domain $\mathfrak{E}$ on $\mathfrak{F}$ is mapped *conformally* onto the unit disc E, $|z| < 1$, of the complex z-plane is equivalent to the following. A one-to-one map of E onto $\mathfrak{E}$ is given with the property: for each fixed z_1 in E, the difference $z - z_1$ is a local parameter at the point $\mathfrak{p}_1$ of $\mathfrak{E}$ corresponding to z_1. If a and b are positive numbers, $b < a < 1$, then the inequalities $|z| \leq a$ or $b \leq |z| \leq a$ define a *z-disc* K and a *z-ring* R, respectively. Let ψ be a continuous density on $\mathfrak{F}$, which is expressed by a function $u(x, y)$ in $\mathfrak{E}$. In § 10 two facts were proved.

(1) If ψ vanishes outside of the z-ring R, then the integral of ψ over $\mathfrak{F}$ is equal to the integral over R,

$$\int_R \psi = 2\pi \int_b^a \int_0^1 u \; d\phi \; r \; dr.$$

Here u stands for $u(r \text{ co } \phi, r \text{ si } \phi)$; for the full disc K, set $b = 0$.

(2) If ψ is nonnegative, then $\int_{\mathfrak{F}} \psi \geq \int_R \psi$.

We write the Schwarz inequality again for arbitrary continuous functions u and v in K:

$$\Big(\int_K uv\Big)^2 \leq \int_K u^2 \cdot \int_K v^2.$$

From this follows the "*rule of quadratic composition*": if $\int_K u^2 \leq A^2$, $\int_K v^2 \leq B^2$, Then $(A \geq 0, B \geq 0)$

$$\int_K (u + v)^2 \leq (A + B)^2. \tag{12.3}$$

We begin by studying the connection between the Dirichlet integral and harmonic functions in the *unit disc of the complex z-plane*. For a function $v(z)$ which is continuously differentiable in the interior of this disc, we define the Dirichlet integral $D(v)$ as an improper integral as follows. Integrate first over the disc $|z| \leq a$ of radius $a < 1$,

$$D^a(v) = 2\pi \int_0^a \int_0^1 (\operatorname{grad} v)^2 \, d\phi \, r \, dr, \tag{12.4}$$

and then let a tend to 1. Since $D^a(v)$ increases as a increases, the limit $D(v)$ exists if $D^a(v)$ remains bounded as $a \to 1$. As *admissible functions* we shall allow *those functions which are defined and continuous in the closed disc* $|z| \leq 1$, *continuously differentiable in the interior* $|z| < 1$, *and have a finite Dirichlet integral.* We denote the boundary values of v, $v(\mathrm{ex}\,\phi)$, by $v(\phi)$. Besides the Dirichlet integral (12.4) over the full disc, the ring integral $D_b^a(v)$, in which r is integrated only over the interval $b \leq r \leq a$ $(0 < b < a < 1)$, will also arise. Now we prove the following theorems.

(1) *If the admissible function* $v = u$ *is also harmonic in the interior of the unit disc, then the Poisson equation*

$$u(z) = \int_0^1 \frac{1 - r^2}{1 + r^2 - 2r\,\mathrm{co}(\phi - \theta)} u(\theta)\, d\theta \tag{12.5}$$

is valid.

(2) *If* $u(\theta)$ *is an arbitrary continuous function of the real variable* θ, *with period* 1, *then the function* $u(z)$ *defined by* (12.5) *is a harmonic function in the unit disc and assumes the boundary values* $u(\theta)$.

(3) *If* $v(z)$ *is an admissible function with the boundary values* $v(\theta)$, *then the harmonic function* $u(z)$, *defined by the Poisson integral with boundary values* $v(\theta)$, *has a Dirichlet integral* $D(u) \leq D(v)$; *more precisely,*

$$D(v - u) = D(v) - D(u). \tag{12.6}$$

Proof of (1). Since the open disc is simply connected, the harmonic function u has a uniform conjugate u' in this domain. From

$$\frac{\partial u}{\partial r} = \frac{1}{r}\frac{\partial u'}{\partial \phi} \qquad \text{for} \qquad 0 < r < 1$$

it follows immediately that the derivative with respect to r of the mean value $\int_0^1 u \, d\phi$ on the circle of radius r is zero; hence this mean value is a constant in the whole interval $0 < r < 1$. By letting $r \to 0$ and $r \to 1$ we obtain the well-known fact that the value of u at the center of the unit disc is equal to its mean value $\int_0^1 u \, d\phi$ on the boundary of the unit disc.

If $z = \alpha$ is a value in the open unit disc, then

$$z^* = \frac{z - \alpha}{1 - \bar{\alpha}z}, \qquad z = \frac{z^* + \alpha}{1 + \bar{\alpha}z^*} \tag{12.7}$$

is a one-to-one conformal transformation of the unit disc onto itself and carries the point $z = \alpha$ into the origin $z^* = 0$. Let $\bar{z}$ denote here (as always if no other definition is given) the conjugate $x - iy$ of the complex number $z = x + iy$. Set $z = r \operatorname{ex} \phi$, $z^* = r^* \operatorname{ex} \phi^*$. Since u is a harmonic function of z^*, we have

$$u(z^* = 0) = \int_0^1 u \, d\phi^* \qquad (r^* = 1). \tag{12.8}$$

From (12.7) it follows that

$$\frac{dz^*}{z^*} = \left(\frac{1}{z - \alpha} + \frac{\bar{\alpha}}{1 - \bar{\alpha}z}\right) dz = \frac{1 - |\alpha|^2}{(z - \alpha)(1 - \bar{\alpha}z)} dz ;$$

hence on the unit circle, where $z\bar{z} = 1$,

$$\frac{dz^*}{z^*} = \frac{dz}{z} \frac{1 - |\alpha|^2}{(1 - \alpha\bar{z})(1 - \bar{\alpha}z)}$$

or

$$d\phi^* = d\phi \frac{1 - |\alpha|^2}{|1 - \alpha\bar{z}|^2}.$$

Thus from (12.8)

$$u(\alpha) = \int_0^1 \frac{1 - |\alpha|^2}{|1 - \alpha \cdot \operatorname{ex}(-\phi)|^2} u(\phi) \, d\phi ,$$

and this becomes the Poison equation (12.5) if α is replaced by z and the variable of integration ϕ is replaced by θ.

Proof of (2). That the function (12.5), where $u(\theta)$ is a given continuous function of the real variable θ with period 1, is a harmonic function in the open unit disc, follows simply from the fact that it is the real part of the analytic function

$$f(z) = \int_0^1 \frac{\zeta + z}{\zeta - z} u(\theta) \, d\theta \qquad (\zeta = \operatorname{ex} \theta).$$

That it assumes the given continuous boundary values u was first proved by

H. A. Schwarz.[5] The proof is found in all textbooks on function theory. It depends on the following facts about the "kernel"

$$\frac{1-r^2}{1+r^2-2r\,\mathrm{co}\,\phi} \qquad (0 \leq r < 1)$$

of the integral representation (12.5): (i) it is always positive; (ii) its integral over the full interval $-\frac{1}{2} \leq \phi \leq \frac{1}{2}$ is equal to 1; (iii) if an arbitrarily small part of this interval about $\phi = 0$ is deleted, then on the remainder, that is,

$$-\tfrac{1}{2} \leq \phi \leq -\varepsilon \quad \text{and} \quad \varepsilon \leq \phi \leq \tfrac{1}{2} \qquad (\varepsilon > 0)$$

it tends to 0 uniformly as $r \uparrow 1$.

For $|z| < 1$, $\zeta = \mathrm{ex}\,\theta$,

$$\frac{\zeta + z}{\zeta - z} = \frac{2\zeta}{\zeta - z} - 1 = 1 + 2\sum_{n=1}^{\infty} z^n \,\mathrm{ex}(-n\theta).$$

Therefore the analytic function $f(z)$, whose real part is the harmonic function u with the boundary values $u(\theta) = v(\theta)$, may be expressed by the absolutely convergent power series

$$f(z) = \sum_{n=0}^{\infty} c_n z^n$$

in the open unit disc, where

$$c_0 = \int_0^1 v(\theta)\,d\theta, \qquad c_n = 2\int_0^1 v(\theta)\,\mathrm{ex}(-n\theta)\,d\theta \qquad (n \geq 1).$$

Thus the real part has the development

$$u(z) = a_0 + \sum_{n=1}^{\infty} (a_n r^n \,\mathrm{co}(n\phi) + b_n r^n \,\mathrm{si}(n\phi)) \tag{12.9}$$

with the coefficients

$$a_0 = \int_0^1 v(\theta)\,d\theta, \qquad a_n = 2\int_0^1 v(\theta)\,\mathrm{co}(n\theta)\,d\theta, \qquad b_n = 2\int_0^1 v(\theta)\,\mathrm{si}(n\theta)\,d\theta.$$

The series for $f(z)$ is uniformly convergent in any disc $|z| \leq q$ of radius $q < 1$. We express this by speaking of "*uniform convergence except at the boundary.*" This sort of convergence holds not only for the series for $f(z)$, but also for the series for any derivative of $f(z)$ obtained from term by term differentiation. From this fact one derives the following convergence theorem.

[5]) *Gesammelte Abhandlungen*, **2**, 186–198.

Let $u_1, u_2, \ldots$ be a sequence of harmonic functions in the open unit disc which converges uniformly, except at the boundary, to a limit function u. Then u is also a potential function; also all derivatives of u_n converge uniformly, except at the boundary, to the corresponding derivative of u.

To prove this, express u_n by the Poisson integral over the circle of radius $a < 1$, and use these expressions in a disc $|z| \leq a' < a$.

Proof of (3). Attempts were made to prove the third theorem as follows. Apply Green's formula and the fact that u is harmonic to turn $D(w, u)$, $w = v - u$, into the integral $\int w(\partial u/\partial r)\, d\phi$ over the boundary of the unit circle. Observe that this integral vanishes since $w = 0$ on the boundary. It is clear that the argument in this form will not hold water because of the unpredictable behavior of the normal derivative $\partial u/\partial r$ under approach to the boundary of the unit disc. One must keep this in mind to evaluate properly the following proof which stems from Hadamard and Zaremba.[6]

With the notations

$$P_n = \frac{1}{\sqrt{\pi n}} r^n \operatorname{co}(n\phi), \qquad Q_n = \frac{1}{\sqrt{\pi n}} r^n \operatorname{si}(n\phi),$$

$$A_n = \sqrt{\pi n}\, a_n, \qquad B_n = \sqrt{\pi n}\, b_n \qquad (n = 1, 2, \ldots)$$

we write the development (12.9), which converges uniformly, together with all its derivatives, in an arbitrary disc $K_q : |z| \leq q$ of radius $q < 1$, in the form

$$u - a_0 = \sum_{n=1}^{\infty} \{A_n P_n(xy) + B_n Q_n(xy)\}. \tag{12.10}$$

Here A_n and B_n may be expressed as surface integrals,

$$A_n = D(v, P_n), \qquad B_n = D(v, Q_n). \tag{12.11}$$

For since Q_n is the conjugate of P_n, it follows from the Stokes' = Green's formula that the integral of (grad v, grad P_n) = [grad v, grad Q_n] over the disc $K_q : 0 \leq r \leq q$, $0 \leq \phi \leq 1$, is equal to the integral

$$\int v\, dQ_n = \frac{2\pi n}{\sqrt{\pi n}} q^n \int_0^1 v(q \operatorname{ex} \phi) \operatorname{co}(n\phi)\, d\phi$$

---

[6]) J. Hadamard, *Sur le principe de Dirichlet*, Bull. Soc. math. France, **34** (1906) 135–139. S. Zaremba, loc. cit., 206ff.

over the circumference $r = q$. From the limit $q \to 1$ we obtain

$$D(v, P_n) = \sqrt{\pi n} a_n = A_n.$$

Analogous equations hold with Q_n in place of P_n.

By setting $v = P_n$ or Q_n in these equations one obtains the "orthogonality relations"

$$\text{(I)} \qquad \begin{cases} D^q(P_n, Q_m) = 0 & \text{without exception,} \\ D^q(P_n, P_m) = 0, \quad D^q(Q_n, Q_m) = 0 & \text{except for } n = m, \\ D^q(P_n) = D^q(Q_n) = q^{2n} & (m, n = 1, 2, \ldots). \end{cases}$$

For

$$J^q(u, v) = 2\pi \int_0^q \int_0^1 uv \, d\phi \, r \, dr$$

the following orthogonality relations hold:

$$\text{(II)} \qquad \begin{cases} J^q(P_n, Q_m) = 0 & \text{without exception,} \\ J^q(P_n, P_m) = 0, \quad J^q(Q_n, Q_m) = 0 & \text{except for } n = m, \\ J^q(P_n) = J^q(Q_n) = \dfrac{q^{2(n+1)}}{2n(n+1)}. \end{cases}$$

One can compute $D^q(u)$ from the series (12.10) by termwise integration; using the equations (I), the result is

$$D^q(u) = \sum_{n=1}^{\infty} q^{2n}(A_n^2 + B_n^2). \tag{12.12}$$

On the other hand, it follows from the same formulas and the expression (12.11) that

$$D\Big(v - \sum_{\nu=1}^{n} [A_\nu P_\nu + B_\nu Q_\nu]\Big) = D(v) - \sum_{\nu=1}^{n} (A_\nu^2 + B_\nu^2).$$

Therefore the series

$$\sum_{n=1}^{\infty} (A_n^2 + B_n^2)$$

must remain $\leq D(v)$, no matter how far one sums it; hence it is convergent. Combining this result with the equation (12.12) we have

$$D^q(u) \leq D(v),$$

and, letting $q \to 1$, it follows that $D(u)$ exists and

$$D(u) \leq D(v). \tag{12.13}$$

Among all admissible functions which assume the same boundary values as v, the potential function minimizes the Dirichlet integral.

It follows from (12.12), by letting $q \to 1$, that

$$D(u) = \sum_{n=1}^{\infty} (A_n^2 + B_n^2). \tag{12.14}$$

For, on the one hand,

$$D^q(u) \leq \sum_{n=1}^{\infty} (A_n^2 + B_n^2), \qquad \text{thus} \qquad D(u) \leq \sum_{n=1}^{\infty} (A_n^2 + B_n^2);$$

on the other hand, the inequality

$$D^q(u) \geq \sum_{\nu=1}^{n} (A_\nu^2 + B_\nu^2) q^{2\nu}$$

implies the inequality

$$D(u) \geq \sum_{\nu=1}^{n} (A_\nu^2 + B_\nu^2)$$

for each n.

The equation (12.6), which sharpens the inequality (12.13), may be derived as follows. Set $v - u = w$ and compute $D^q(w, u)$ by using the series (12.10) for u. (We write only the first part of it, denoting the second by ●.) Thus one obtains $\Sigma_n A_n D_0^q(w, P_n) +$ ● . But since

$$D(w, P_n) = 2\sqrt{\pi n} \int_0^1 \{v(\phi) - u(\phi)\} \operatorname{co} n\phi \, d\phi = 0,$$

one can replace D_0^q by $- D_q^1$. Then from

$$D^q(w, u) = -\sum_n A_n D_q^1(w, P_n) - \bullet$$

it follows by the Schwarz inequality that[7]

$$\{D^q(w, u)\}^2 \leq D_q^1(w) \cdot \sum_n (A_n^2 + \bullet) = D_q^1(w) D(u).$$

---

[7]) The Schwarz inequality gives

$$\{D^q(w, u)\}^2 \leq \{\sum_n (s_n^2 + t_n^2)\} \{\sum_n (A_n^2 + B_n^2)\},$$

where $s_n = D_q^1(w, P_n)$ and $t_n = D_q^1(w, Q_n)$. The rest follows from the Bessel inequality corresponding to the annulus $(q, 1)$:

$$\sum_1^{\infty} (s_n^2 + t_n^2) \leq \sum_1^{\infty} \frac{s_n^2 + t_n^2}{1 - q^{2n}} \leq D_q^1(w).$$

Letting $q \to 1$, we see that the first variation $D(w, u)$ vanishes, and hence

$$D(v) = D(u + w) = D(u) + D(w).$$

Another way of proving that the first variation vanishes is by means of the inequality $D(u + \varepsilon w) \geq D(u)$, which is valid for every real ε.

With this, the three facts stated above are proved. From

$$|c_1|^2 = a_1^2 + b_1^2 = \frac{1}{\pi}(A_1^2 + B_1^2) \leq \frac{1}{\pi} D(u)$$

we derive the following inequality for u and the analytic function $f(z)$ with real part u,

$$\left|\frac{df}{dz}\right|^2 = \left(\frac{\partial u}{\partial x}\right)^2 + \left(\frac{\partial u}{\partial y}\right)^2 \leq \frac{1}{\pi} D(u) \leq \frac{1}{\pi} D(v), \tag{12.15}$$

which is valid for $z = 0$. Similarly,

$$\left|\frac{1}{n!}\frac{d^n f}{dz^n}\right|^2_{z=0} \leq \frac{1}{\pi n} D(u).$$

We shall extend the inequality (12.15) from the center to an arbitrary point $z = \alpha$ of the open unit disc by means of the transformation (12.7) which leaves the Dirichlet integral unchanged. Now

$$\left(\frac{df}{dz}\right)_{z=\alpha} = \left(\frac{df}{dz^*}\bigg/\frac{dz}{dz^*}\right)_{z^*=0},$$

and, by the little computation above, dz/dz^* has the value $1 - |\alpha|^2$ at the point $z^* = 0$. Then from $|df/dz^*|^2_{z^*=0} \leq D(u)/\pi$ we get

$$\left|\frac{df}{dz}\right|^2 = \left(\frac{\partial u}{\partial x}\right)^2 + \left(\frac{\partial u}{\partial y}\right)^2 \leq \frac{D(u)}{\pi(1 - r^2)^2}. \tag{12.16}$$

Thus one finds that the derivatives $\partial u/\partial x$ and $\partial u/\partial y$ become infinite in at most a first-order manner as $|z| = r \to 1$; we have a precise estimate for this growth. We shall make use of this fact shortly.

But first another lemma, concerning the approximation of an admissible function in the unit disc by a constant. Since v is constant if $(\text{grad } v)^2 = 0$, one would expect that an inequality of the form

$$J^1(v - a) = J(v - a) \leq \text{Const}\, D(v)$$

would hold for an appropriate choice of the constant a. We shall show that this is in fact so if we pick a to be the mean value of v on the periphery of the unit disc; the constant factor will be $\frac{1}{2}$. In a fashion similar to the derivation of (12.14) from the formulas (I) via (12.12), we obtain from the formulas (II) the equation

$$J(u-a) = \sum_{n=1}^{\infty} \frac{A_n^2 + B_n^2}{2n(n+1)}$$

for the harmonic function u. Combining this with (12.14) we obtain the inequality

$$J(u-a) \leq \tfrac{1}{4} D(u). \tag{12.17}$$

Thus we have an inequality of the desired type for a *potential function* u. We split the arbitrary admissible function v into the harmonic function u with the same boundary values and *an admissible function* w *which vanishes on the boundary*. I claim that an inequality similar to (12.17) is also valid for this last function:

$$J(w) \leq \tfrac{1}{4} D(w). \tag{12.18}$$

Combining this with (12.17), one gets

$$\begin{aligned} J(v-a) &= J((u-a)+w) \leq 2\{J(u-a) + J(w)\} \\ &\leq \tfrac{1}{2}\{D(u) + D(w)\} = \tfrac{1}{2} D(v), \end{aligned}$$

and the proof of the inequality

$$J(v-a) \leq \tfrac{1}{2} D(v) \tag{12.19}$$

is finished.

We derive (12.18) as follows. Since w vanishes on the circumference,

$$w(r \cdot \operatorname{ex} \phi) = \int_1^r \frac{\partial w(\rho \cdot \operatorname{ex} \phi)}{\partial \rho} \, d\rho.$$

If we write the integrand in the form $(\partial w/\partial \rho)\sqrt{\rho}\,(1/\sqrt{\rho})$, then the Schwarz inequality yields

$$\{w(r \cdot \operatorname{ex} \phi)\}^2 \leq \int_r^1 \left(\frac{\partial w}{\partial \rho}\right)^2 \rho \, d\rho \int_r^1 \frac{d\rho}{\rho},$$

and, integrating with respect to ϕ:

$$2\pi \int_0^1 w^2(r \cdot \operatorname{ex} \phi) \, d\phi \leq \log \frac{1}{r} D_r^1(w). \tag{12.20}$$

We replace $D_r^1(w)$ by $D(w)$, which is larger, multiply by $r\,dr$, and integrate from $r = 0$ to $r = 1$; we get in fact then

$$J(w) \leq \int_0^1 r \cdot \log \frac{1}{r}\, dr\, D(w) = \tfrac{1}{4} D(w).$$

After these investigations, concerned with harmonic functions in the unit disc, we return to arbitrary Riemann surfaces. Let a domain $\mathfrak{G}$ thereon be mapped one-to-one and conformally onto an open disc, centered at the origin of the z-plane, which contains the closed disc $E: |z| \leq a$ ("z-disc"). Again we take $a = 1$ (which means that the image of $\mathfrak{G}$ has a radius greater than 1). Now given a continuously differentiable function on the Riemann surface, we wish to modify it in the region of the disc E so as to decrease the Dirichlet integral as much as possible. We already know what the best solution is: we replace $v(z)$ in the interior of E by that harmonic function $u(z)$ which has the same boundary values. But the continuous differentiability on the boundary of E is (almost certainly) lost in the process. Therefore we must *smooth* the function there in a suitable way.

For this purpose we choose a positive number $q < 1$ (with the intention of letting q tend to 1 later) and set up the following function for $q \leq r \leq 1$:

$$\chi(r) = \frac{1}{2} + \frac{1}{2} \mathrm{co} \frac{\pi(1-r)}{1-q}. \tag{12.21}$$

It varies between 0 and 1, $\chi(r)$ has a second-order zero at $r = q$, and $1 - \chi(r)$ has a second-order zero at $r = 1$. Set $\chi(r) = 0$ for $r < q$. If we set

$$\tilde{v} = u + \chi(r)\, w = v - \{1 - \chi(r)\}\, w \qquad (w = v - u), \tag{12.22}$$

then $\tilde{v}$ is continuously differentiable across both circles $r = 1$ and $r = q$. Here we use the fact that the derivatives of u become infinite in at most a first-order manner as $r \to 1$. We claim that $D(\tilde{v}) - D(u) = D(\tilde{v} - u)$ may be made as small as we wish by choosing q sufficiently close to 1. Since

$$\frac{\partial}{\partial x}\{\chi(r) w\} = \chi(r) \frac{\partial w}{\partial x} + w \frac{d\chi}{dr} \frac{x}{r},$$

the contribution of the first term to the Dirichlet integral D_q^1 of $\chi(r) w$ is

$$2\pi \int_q^1 \int_0^1 \chi^2(r) (\operatorname{grad} w)^2\, d\phi\, r\, dr,$$

which is $\leq D_q^1(w)$. The contribution of the second term is

$$2\pi \int_q^1 \int_0^1 w^2 \left(\frac{d\chi}{dr}\right)^2 d\phi\, r\, dr.$$

If we set $\pi(1 - r)/(1 - q) = s$ and observe the inequality (12.20) which holds in the interval $q \leq r \leq 1$

$$2\pi \int_0^1 w^2 d\phi \leq \left(\frac{1}{r} - 1\right) D_r^1(w) \leq \left(\frac{1}{r} - 1\right) D_q^1(w),$$

then we obtain the upper bound

$$D_q^1(w) \tfrac{1}{\pi} \int_0^\pi \sin^2 s \, s \, ds$$

for the contribution of the second term. The constant factor may be evaluated immediately by writing the integrand in the form $\frac{1}{2}(1 - \cos 2s)$; one obtains $(\pi/4)^2$. According to the rule of (12.3) of "quadratic composition" we get

$$D(\tilde{v}) - D(u) = D(\tilde{v} - u) \leq \left(1 + \frac{\pi}{4}\right)^2 D_q^1(w).$$

In fact, the right-hand side tends to 0 as $q \to 1$. If v is not harmonic in the disc E, then it may be replaced by a $\tilde{v}$, which is still continuously differentiable across $|z| = 1$, whose Dirichlet integral over the disc E is smaller than that of v. [If v is harmonic in E, then u and $\tilde{v}$ coincide with v; then of course $D(\tilde{v}) = D(v)$.]

§ 13. Scheme for the construction of the potential arising from a doublet source

Instead of constructing an analytic function on the given closed Riemann Surface $\mathfrak{F}$, we first look for only the real part of such a function; that is, a real harmonic function on the surface. But (except for the constants) there are no potential functions which are regular everywhere on the surface. We shall therefore permit the harmonic function to have a singularity at one point, and that singularity will be the simplest that there is: the harmonic function will behave there like the real part of an analytic function which has a pole of order one at the point. Thus we set out to prove the following existence theorem.

Let $\mathfrak{o}$ be an arbitrary point on $\mathfrak{F}$ and let $\zeta = \xi + i\eta = \rho \operatorname{ex} \psi$ be a local parameter at $\mathfrak{o}$. Then there exists a potential function U on $\mathfrak{F}$ which is regular at all points except $\mathfrak{o}$ and which, in a neighborhood of $\mathfrak{o}$, differs from $\xi/(\xi^2 + \eta^2)$ by a potential function which is regular at $\mathfrak{o}$.

Clearly these properties determine U uniquely to within an additive constant. Even if $\mathfrak{F}$ is a Riemann surface in the abstract sense, as introduced in Chapter I, instead of an actual surface in space, it is permissible to regard U as the potential of an incompressible stationary irrotational fluid flow on $\mathfrak{F}$, a flow with no sources except $\mathfrak{o}$; at the point $\mathfrak{o}$ there is a "doublet source" of moment 2π, whose direction is that of the positive ξ-axis. It certainly is somewhat daring to infer the existence of U from its hydrodynamic significance.[8]

If U had a Dirichlet integral, then it would be called the energy of the fluid flow. Unfortunately, the integral is infinite, because of the behavior of U at the point $\mathfrak{o}$. Therefore we proceed as follows. To the local parameter ζ at $\mathfrak{o}$ there corresponds a domain $\mathfrak{G}$, containing $\mathfrak{o}$, that is mapped by ζ one-to-one and conformally onto an open disc, with $\mathfrak{o}$ mapping into the center. Let the ζ-disc $K_0 : | \zeta | \leq a_0$ be contained in the image of $\mathfrak{G}$. There is also a ζ-disc $K'_0 : | \zeta | \leq b_0$, of somewhat larger radius, contained in that neighborhood $\mathfrak{G}$. We form the harmonic function

$$\Phi = \frac{\xi}{\xi^2 + \eta^2} + \frac{\xi}{a_0^2} \tag{13.1}$$

in $\mathfrak{G}$. It has the required singularity and also has the property that its normal derivative vanishes on the periphery of K_0, which follows from

$$\Phi = \frac{\mathrm{co}\,\psi}{\rho} + \frac{\rho\,\mathrm{co}\,\psi}{a_0^2},$$

$$\frac{\partial \Phi}{\partial \rho} = -\frac{\mathrm{co}\,\psi}{\rho^2} + \frac{\mathrm{co}\,\psi}{a_0^2} = 0 \qquad \text{for} \qquad \rho = a_0 .$$

We now consider

$$u = \begin{cases} U \text{ on } \mathfrak{F}, \text{ except in } K_0 \\ U - \Phi \text{ in } K_0 . \end{cases} \tag{13.2}$$

Although this function has a jump along the periphery κ of K_0, one can still form the Dirichlet integral $D(u)$, as we shall explain immediately.

Among all functions v which have the same jump as u along κ, but which are otherwise continuously differentiable, u minimizes the Dirichlet integral $D(v)$. We shall characterize the functions v, which are being compared with u, more precisely. I call $\mathfrak{F} - K_0$ the *punched surface*, K_0 the *hole*, K'_0 the *lid*, and the annulus $a_0 \leq | \zeta | \leq b_0$, in which the lid overlaps the punched surface, the *lock-ring*. *A function v will be admitted to competition then if it is continu-*

[8]) The presentation in Klein's monograph of 1882, already cited, is based on these and similar physical considerations. The diagrams of streams in that monograph are very instructive.

ously differentiable on the punched surface, if on the hole it coincides with a function v^* which is continuously differentiable on the lid, and if $v = v^* + \Phi$ on the lock-ring. If $\mathfrak{F}$ is noncompact, we also demand that $D(v)$ be finite. If v is the symbol for any competing function, then v^* will always denote the continuously differentiable function on the lid that we have just characterized. The difference of two competing functions is continuously differentiable on all of $\mathfrak{F}$.

The Dirichlet integral $D(v)$ consists of two parts, the integral of $(\operatorname{grad} v)^2$ over the punched surface, and the integral of $(\operatorname{grad} v^*)^2$, to be computed in polar coordinates ρ, ψ, over the hole. According to the method given in Chapter I, § 10, for integrating over a punched surface, one chooses a continuous smoothing function λ for the lock-ring (which is hence $= 1$ in the hole, vanishes outside the lid, and in the ring is a function of ρ which decreases from 1 to 0) and forms the density

$$\Psi(v) = (1 - \lambda)(\operatorname{grad} v)^2 + \lambda(\operatorname{grad} v^*)^2 , \tag{13.3}$$

which is continuous on all of $\mathfrak{F}$. To the integral $D_\lambda(v)$ of this density there is added the integral $D'_\lambda(v)$, over the lock-ring, of

$$\Psi'(v) = \lambda\{(\operatorname{grad} v)^2 - (\operatorname{grad} v^*)^2\} . \tag{13.4}$$

The result, $D(v)$, is independent of the choice of the smoothing function λ. That competing functions exist is trivial. We obtain one, which vanishes outside the hole, by choosing for v in the hole $-2\,\xi/a_0^2$, which has the same boundary values as $-\Phi$. For this has the prescribed jump Φ over the boundary of the hole. For its Dirichlet integral one finds immediately the value $4\,\pi/a_0^2$, and this number is an upper bound for the minimum sought. To be sure, v must be smoothed in the lock-ring so that it will satisfy our admissibility conditions. With a positive number $\delta < b_0 - a_0$ one forms the function χ, which is 1 for $\rho \leq a_0$, which vanishes outside the disc $\rho < a_0 + \delta$, and is given in the interval $a_0 \leq \rho \leq a_0 + \delta$ by the formula

$$\chi(\rho) = \frac{1}{2} + \frac{1}{2}\cos\frac{\pi(\rho - a_0)}{\delta}$$

already familiar to us from (12.21). Instead of $v = 0$ on the punched surface and $v^* = -2\,\xi/a_0^2$ on the hole, one now sets

$$\begin{aligned} v &= \chi\left(\frac{\xi}{\xi^2 + \eta^2} - \frac{\xi}{a_0^2}\right) && \text{on the punched surface}, \\ v^* &= (\chi - 1)\frac{\xi}{\xi^2 + \eta^2} - (1 + \chi)\frac{\xi}{a_0^2} && \text{on the lid}. \end{aligned} \tag{13.5}$$

For the admissible function v (which, to be sure, does not vanish outside the hole, but does vanish outside the lid) the value of the Dirichlet integral is the number $4\pi/a_0^2$, computed above, plus the value of the integral of $(\text{grad}\, v)^2$ over the ring $a_0 \leq \rho \leq a_0 + \delta$, where v is given by (13.5). This added term is of the order of magnitude of δ and hence tends to 0 with δ. In this sense it is legitimate to use the simpler function v, which vanishes outside the hole and $= -2\xi/a_0^2$ in the hole, as a comparison function.

To justify this attack on our problem as a minimization problem, we shall show next that if the minimizing function u exists (it is then itself a function admitted to the competition), then it has the desired properties; that is, u is regular harmonic on the punched surface and the associated u^* is regular harmonic on the lid.

If $\mathfrak{p}_0$ is an interior point of the punched surface or of the lid and if z is a local parameter at $\mathfrak{p}_0$, then we can find a closed z-disc K, $|z| \leq a$, which is contained in the interior of the punched surface or of the lid. Let v_1 and v_2 be two competing functions which coincide outside the disc K. That is, $v_1 - v_2 = 0$, or $v_1^* - v_2^* = 0$ outside K, according to whether K lies in the punched surface or in the lid. I maintain that in the first case

$$D(v_1) - D(v_2) = D_K(v_1) - D_K(v_2), \tag{13.6}$$

and

$$D(v_1) - D(v_2) = D_K(v_1^*) - D_K(v_2^*) \tag{13.6*}$$

in the second case. Here $D_K(v)$ is the Dirichlet integral over the disc K, which is to be computed with the associated polar coordinates. Once this result is established, then it obviously follows from the fact proved in the preceding paragraph that by modifying v inside a disc one can depress the value of $D(v)$, except in the case that v is harmonic in K.

The equation (13.6) follows immediately if one chooses the smoothing function λ in the definition of $D(v)$ in (13.3) and (13.4) so that it vanishes in all of K; this is possible even if K penetrates into the lock-ring. On the other hand, in the proof of (13.6*) one will choose λ so that it is identically 1 in K; also here one uses $\zeta - \zeta(\mathfrak{p}_0)$ as the local parameter z. Then since the continuous density

$$(1-\lambda)\{(\text{grad}\, v_1)^2 - (\text{grad}\, v_2)^2\} + \lambda\{(\text{grad}\, v_1^*)^2 - (\text{grad}\, v_2^*)^2\}$$

vanishes outside K and is equal to $(\text{grad}\, v_1^*)^2 - (\text{grad}\, v_2^*)^2$ in K, the integral of $\Psi(v_1) - \Psi(v_2)$ over all of K contributes $D_K(v_1^*) - D_K(v_2^*)$ to $D(v_1) - D(v_2)$. The additional contribution

$$4\pi \int_{a_0}^{b_0} \int_0^1 (\text{grad}\,(v_1^* - v_2^*), \text{grad}\, \Phi)\, d\psi\, \rho\, d\rho$$

results from (13.4). By Green's formula this is equal to the difference of the integrals over the boundary:

$$4\pi\left[\int_0^1 (\overset{*}{v}_1 - \overset{*}{v}_2)\frac{\partial \Phi}{\partial \rho}\rho\, d\psi\right]_{\varrho=a_0}^{\varrho=b_0}.$$

But this is zero since $v_1 - v_2 = 0$ on the outer boundary of the lock-ring and $\partial\Phi/\partial\rho = 0$ on the inner one. Here the fact that by adding the term ξ/a_0^2 to the singularity $\xi/(\xi^2 + \eta^2)$ a Φ was formed with vanishing normal derivative on the boundary of the hole has become important.

The values of the Dirichlet integral $D(v)$ for all possible admissible functions v have a definite infimum $d\,(\leq 4\,\pi/a_0^2)$. Our task is to show that this infimum is attained by a certain admissible function u.

We can handle the minimum problem as a problem in *orthogonal projection* if we regard an admissible function v as a vector, in an infinite-dimensional space, of length $\sqrt{D(v)}$. The functions w which are continuously differentiable on all of $\mathfrak{F}$ and have a finite Dirichlet integral form a linear manifold $\mathfrak{W}$ in this vector space. Every admissible function may be obtained from one fixed v by subtracting an arbitrary element w of $\mathfrak{W}$ from v. Thus one should choose the vector w in $\mathfrak{W}$ such that the length of $v - w = u$ is as small as possible. But that is the problem of orthogonal projection: w is the orthogonal projection of v on $\mathfrak{W}$, u the perpendicular. Using this formulation, one obtains in a natural fashion, from the theorem that the length of the difference of two vectors cannot exceed the sum of the lengths, the following inequality, first derived by Beppo Levi.[9] For any two admissible functions v_1 and v_2,

$$\sqrt{D(v_1 - v_2)} \leq \sqrt{D(v_1) - d} + \sqrt{D(v_2) - d}. \tag{13.7}$$

Without using the existence of the "perpendicular," that is, the minimal function u, one can prove this very simply as follows.

If λ_1 and λ_2 are any two real constants, $\lambda_1 + \lambda_2 \neq 0$, and if v_1 and v_2 are competing functions, then so is

$$\frac{\lambda_1 v_1 + \lambda_2 v_2}{\lambda_1 + \lambda_2}.$$

Hence

$$D(\lambda_1 v_1 + \lambda_2 v_2) \geq d(\lambda_1 + \lambda_2)^2.$$

---

[9]) *Sul principio di Dirichlet,* Rend. Circ. Mat. Palermo, **22** (1908) 293–360, §7. For the method under discussion see H. Weyl, *The method of orthogonal projection in potential theory,* Duke Math. Jour., **7** (1940) 411–444.

This inequality remains valid for $\lambda_1 + \lambda_2 = 0$. Thus the quadratic form

$$\lambda_1^2[D(v_1) - d] + 2\lambda_1\lambda_2[D(v_1v_2) - d] + \lambda_2^2[D(v_2) - d]$$

in λ_1 and λ_2 is always ≥ 0; therefore

$$[D(v_1) - d][D(v_2) - d] \geq [D(v_1v_2) - d]^2.$$

Then

$$\begin{aligned}
0 \leq D(v_1 - v_2) &= D(v_1) - 2D(v_1v_2) + D(v_2)\\
&= [D(v_1) - d] + [D(v_2) - d] - 2[D(v_1v_2) - d]\\
&\leq [D(v_1) - d] + [D(v_2) - d] + 2\sqrt{[D(v_1) - d][D(v_2) - d]}\\
&= \{\sqrt{D(v_1) - d} + \sqrt{D(v_2) - d}\}^2.
\end{aligned}$$

From Levi's inequality it follows that the difference of two minimal functions must be a constant; that is, *the solution of the minimal problem is unique to within an additive constant.* To normalize this constant, we require that the integral of v^* over the boundary of the hole shall be zero. More precisely, the inequality states that if v_1 and v_2 are two competing functions whose Dirichlet integrals come close to the infimum d, then their difference $w = v_1 - v_2$ has a small Dirichlet integral. From this one can draw a conclusion similar to the following: if one varies v so as to force $D(v)$ to approach the infimum d, then v itself will converge to a limit function, which will be the minimal function sought.

If $\theta(v)$ is a number depending on the competing function v, then a limit equation such as $\lim_v \theta(v) = \theta_0$ is to mean that for each positive number δ, a positive number ε can be given such that $|\theta(v) - \theta_0| < \delta$ whenever $D(v) < d + \varepsilon$. Let $\mathfrak{p}$ be any point of the surface, $z = x + iy$ a local parameter at $\mathfrak{p}$, and $K: |z| \leq a$ a z-disc. Then Schwarz' inequality, applied to the product of the two functions $\partial w/\partial x$ (or $\partial w/\partial y$) and 1 yields

$$\left(\iint_K \frac{\partial w}{\partial x} dx\, dy\right)^2 + \left(\iint_K \frac{\partial w}{\partial y} dx\, dy\right)^2 \leq \pi a^2 D_K(w) \leq \pi a^2 D(w).$$

Thus, if K is contained in the punched surface, the limit

$$\lim_v \iint_K \frac{\partial v}{\partial x} dx\, dy$$

exists. If K is contained in the lid, then v is to be replaced by v^*. (Moreover, this limit exists uniformly for all subregions of K which have Jordan content.) If one cannot derive the convergence of the differential quotients $\partial v/\partial x$, $\partial v/\partial y$ in this way, at least one can derive the convergence of their surface integrals. It is important to get from the derivatives back to the *function v itself*, and we shall show shortly that $\lim_v \iint_K v\,dx\,dy$ exists in the same sense. If the minimal function u is available, this limit will be $\iint_K u\,dx\,dy$; but, since u is harmonic, this is equal to the area πa^2 of K times the value of u at the center $\mathfrak{p}$ of the disc. Thus we arrive at the idea of defining $u(\mathfrak{p})$ by the equation

$$u(\mathfrak{p}) = \frac{2}{a^2} \lim_v \int_0^a \int_0^1 v\, d\phi\; r\, dr\,.$$

Then the last step must be to show that this function has all the desired properties and, in particular, that it is the minimal function sought. The basic idea of the construction is thus the following: *since the limit function will be harmonic, the proposed convergence-creating integration over the disc will have no influence, in the limit, on the value at a point.*

§ 14. The proof

Again let z be a local parameter at the arbitrary point $\mathfrak{p}$ on $\mathfrak{F}$ and let K, $|z| \leq a$, be a z-disc. We shall show that *there exists a number C such that every continuously differentiable function w on $\mathfrak{F}$, with finite Dirichlet integral $D(w)$, which is normalized so that its integral over the boundary of the hole vanishes, satisfies the inequality*

$$J_K(w) = \iint_K w^2\, dx\, dy \leq C D(w). \tag{14.1}$$

For this purpose we form a chain of discs $K_0, K_1, \ldots, K_n$ with centers $\mathfrak{p}_0, \mathfrak{p}_1, \ldots, \mathfrak{p}_n$, beginning with the hole K_0 and ending with the disc K, so that K_i and K_{i+1} overlap (have interior points in common). More precisely, let K_i be a z_i-disc, where z_i is a local parameter at $\mathfrak{p}_i$; let $\mathfrak{p}_0 = \mathfrak{o}$, $\mathfrak{p}_n = \mathfrak{p}$, $z_0 = \zeta$, $z_n = z$. We carry out the transfer step from K_i to K_{i+1} for the first pair $K_0 \to K_1$. By (12.19) and the boundary normalization we have the inequality (14.1) in the form

$$\iint_{K_0} w^2\, dx_0\, dy_0 \leq \frac{a_0^2}{2} D_{K_0}(w) \leq \frac{a_0^2}{2} D(w)$$

for K_0. For K_1 there exists a constant c such that

$$\iint_{K_1} (w - c)^2\, dx_1\, dy_1 \leq \frac{a_1^2}{2} D(w).$$

Let k be a part of the intersection of K_0 and K_1, say a part which is a disc in the z_1-plane of area $\pi(a_1\theta)^2$, $0 < \theta < 1$. In k the variables z_0 and z_1 are connected by a transformation which is regular analytic in k and on the boundary of k. Let m be an upper bound of $| dz_1/dz_0 |$ in k. Then

$$\iint_k w^2 \, dx_1 \, dy_1 \leq m^2 \iint_k w^2 \, dx_0 \, dy_0 \leq \frac{(ma_0)^2}{2} D(w).$$

From this and

$$\iint_k (c - w)^2 \, dx_1 \, dy_1 \leq \frac{a_1^2}{2} D(w) \tag{14.2}$$

it follows from the rule of quadratic composition, applied to the integral over k of the square of the constant $c = (c - w) + w$, that

$$c^2 \pi(a_1 \theta)^2 \leq (a_1 + ma_0)^2 \tfrac{1}{2} D(w).$$

Therefore the integral over K_1 in the z_1-plane of the constant c^2 is at most

$$\left(\frac{a_1 + ma_0}{\theta} \right)^2 \cdot \frac{1}{2} D(w)$$

and hence we obtain, using (14.2), the following estimate of the integral of the square of $w = (w - c) + c$:

$$\iint_{K_1} w^2 \, dx_1 \, dy_1 \leq \left(a_1 + \frac{a_1 + ma_0}{\theta} \right)^2 \frac{1}{2} D(w).$$

Thus in fact we have proved an inequality of the type (14.1) for K_1. Instead of using the disc k, one could, to get as sharp an estimate as possible, have used the whole intersection of K_0 and K_1. But then one would have had to operate with the Jordan content of this domain; we have avoided this in order to use a procedure as elementary as possible.

The step from K_1 to K_2 is similar to that from K_0 to K_1; and so on, until we reach $K_n = K$. This lemma (14.1), which we have just proved, implies that any two admissible functions v_1 and v_2 satisfy the inequality

$$J_K(v_1 - v_2) \leq C \{ \sqrt{D(v_1) - d} + \sqrt{D(v_2) - d} \}^2. \tag{14.3}$$

Let us use the abbreviation $\mathbf{M}v = \mathbf{M}_{z,K}v$ to denote the mean value,

$$\frac{1}{\pi a^2} \iint_K v \, dx \, dy = \frac{2}{a^2} \int_0^a \int_0^1 v \, d\phi \, r \, dr$$

relative to z, of a function v in K. The center of K will be denoted by $\mathfrak{p}_0$ rather than $\mathfrak{p}$. Since by the Schwarz inequality,

$$\{\iint_K (v_1 - v_2)\,dx\,dy\}^2 \leq \iint_K (v_1 - v_2)^2\,dx\,dy\,\pi a^2,$$

it follows from (14.3) that

$$(14.4)\qquad |\mathbf{M}v_1 - \mathbf{M}v_2| \leq \frac{1}{a}\sqrt{\frac{C}{\pi}}\{\sqrt{D(v_1) - d} + \sqrt{D(v_2) - d}\}.$$

Therefore $\lim_v \mathbf{M}v = u(\mathfrak{p}_0)$ exists. If we carry out the limit indicated in $\lim_v$ with v_1 and set $v = v_2$, then (14.4) yields the estimate

$$(14.5)\qquad |u(\mathfrak{p}_0) - \mathbf{M}v| \leq \frac{1}{a}\sqrt{\frac{C}{\pi}}\sqrt{D(v) - d}.$$

This is valid for a disc K in the punched surface. If, on the other hand, K is contained in the lid, then the left-hand side of (14.4) is to be replaced by $\mathbf{M}v_1^* - \mathbf{M}v_2^*$; $u^*(\mathfrak{p}_0)$ is defined as the limit of $\mathbf{M}v^*$, and one has the same estimate as (14.5) for $u^*(\mathfrak{p}_0) - \mathbf{M}v^*$. If for a point of the lock-ring we use a z-disc which is contained in the lock-ring, then we may consider both $\mathbf{M}v$ and $\mathbf{M}v^*$. Their difference is $\mathbf{M}\Phi$, but since Φ is harmonic, $\mathbf{M}\Phi = \Phi(\mathfrak{p}_0)$; therefore the equation $u(\mathfrak{p}_0) = u^*(\mathfrak{p}_0) + \Phi(\mathfrak{p}_0)$ holds for points $\mathfrak{p}_0$ of the lock-ring.

Again let K be contained in the punched surface; after what has been said it is clear what modifications are to be made if K is in the lid. We assume, as will be proved later, that the limit of $\mathbf{M}_{z,K}v$ is independent of the choice of the local parameter z and the z-disc K. For an arbitrary point $\mathfrak{p}$ in K we consider the mean value $\mathbf{M}_\mathfrak{p}$ formed by using the z-disc $k_\mathfrak{p}$, of radius $a - |z(\mathfrak{p})|$, centered at $z(\mathfrak{p})$. The integral of $(v_1 - v_2)^2$ over $k_\mathfrak{p}$ does not exceed the integral over the whole of K. If we determine $u(\mathfrak{p})$ as the limit of $\mathbf{M}_\mathfrak{p}v$, then we get in place of (14.5)

$$|u(\mathfrak{p}) - \mathbf{M}_\mathfrak{p}v| \leq \frac{1}{a - |z(\mathfrak{p})|}\sqrt{\frac{C}{\pi}}\sqrt{D(v) - d}.$$

From this it follows that $\mathbf{M}_\mathfrak{p}v$ converges *uniformly* to $u(\mathfrak{p})$ in any disc concentric with K and of smaller radius.

To prove that $u(\mathfrak{p})$ is a potential function in K, we use the harmonic function $\bar{v}$ in K which coincides with v on the boundary.[10] By the smoothing process given in § 12 we determine an admissible $\tilde{v}$, which coincides with v

[10]) Let it be permitted here that we use the bar to mean something other than the passage from a complex number to its conjugate.

outside K, such that $D_K(\tilde{v})$ comes as close as one wishes to $D_K(\bar{v})$, and in particular such that $D_K(\tilde{v}) \leq D_K(v)$. Then $D(\tilde{v}) \leq D(v)$. The inequality (13.7) remains valid if on the left-hand side (but not on the right!) we replace v_2 by $\tilde{v}_2$; thus

$$\sqrt{D_K(v_1 - \bar{v}_2)} \leq \sqrt{D(v_1) - d} + \sqrt{D(v_2) - d}. \tag{14.6}$$

If one repeats the above argument, with $\bar{v}_2$ in place of v_2, then one obtains the inequality that results from (14.5) by replacing v on the left-hand side by $\bar{v}$. But since $\bar{v}$ is harmonic, $\mathbf{M}_\mathfrak{p}\bar{v} = \bar{v}(\mathfrak{p})$, and therefore $\bar{v}(\mathfrak{p})$ and $u(\mathfrak{p})$ differ absolutely by at most the right-hand side of (14.5). Thus the harmonic function $\bar{v}(\mathfrak{p})$ converges to $u(\mathfrak{p})$ uniformly in any disc concentric with K and of smaller radius. Hence u is harmonic in the interior of K, and not only $\bar{v}$, but also each derivative of $\bar{v}$, converges "uniformly in K except at the boundary" to u or the corresponding derivative of u.

The simple artifice which served us here consists in replacing v in the disc K by the harmonic function $\bar{v}$ before taking the limit $D(v) \to d$; this replacement only depresses the value of $D(v)$. The same artifice must give as a corollary the independence of $\lim_v \mathbf{M}_{z,K} v$ from z and K. The simplest way of formulating this proof is as follows. We take $v_1 = v_2 = v$ in the inequality (14.6):

$$D_K(v - \bar{v}) \leq 4\{D(v) - d\}. \tag{14.7}$$

Since $v - \bar{v}$ vanishes on the boundary of K, we can apply the simple inequality (12.18). The result is

$$J_K(v - \bar{v}) \leq a^2\{D(v) - d\},$$

$$|\mathbf{M}v - \mathbf{M}\bar{v}|^2 \leq |\mathbf{M}v - \bar{v}(\mathfrak{p}_0)|^2 \leq \frac{1}{\pi}\{D(v) - d\},$$

and therefore

$$\lim_v \mathbf{M}_{z,K} v = \lim_v \bar{v}(\mathfrak{p}_0). \tag{14.8}$$

So far we have only repeated the estimation carried out above in a sharpened form for the simpler case $v_1 = v_2$. But now let $z' = x' + iy'$ be another local parameter at $\mathfrak{p}_0$ and let K', $|z'| \leq a'$, be a z'-disc. I assume first that K' is a part of K. In the disc K', $|dz/dz'|$ has a positive lower bound m^{-1}. Therefore

$$\iint_{K'} (v - \bar{v})^2\, dx'\, dy' \leq m^2 \iint_K (v - \bar{v})^2\, dx\, dy,$$

and hence

$$|\mathbf{M}_{z',K'} v - \mathbf{M}_{z',K'} \bar{v}|^2 \leq \left(\frac{ma}{a'}\right)^2 \frac{1}{\pi}\{D(v) - d\}. \tag{14.9}$$

But since $\bar{v}$ is also a harmonic function of x' and y', $\mathbf{M}_{z',K'}\bar{v} = \bar{v}(\mathfrak{p}_0)$. Thus (14.9) furnishes the relation

$$\lim_v \mathop{\mathbf{M}v}_{z',K'} = \lim_v \bar{v}(\mathfrak{p}_0),$$

which, together with (14.8), yields the identity of the limits $\lim_v \mathbf{M}_{z,K}v$ and $\lim_v \mathbf{M}_{z',K'}v$. If K' is not contained in K, then choose a smaller concentric z'-disc, k', which is contained in K. From the result we have just derived it follows that both $\lim_v \mathbf{M}_{z,K}v$ and $\lim_v \mathbf{M}_{z',K'}v$ coincide with $\lim_v \mathbf{M}_{z',k'}v$ and are hence equal.[11]

Thus we have found a harmonic function $u(\mathfrak{p})$, defined on the punched surface, and a harmonic function $u^*(\mathfrak{p})$, defined on the lid, which satisfy the relation $u(\mathfrak{p}) = u^*(\mathfrak{p}) + \Phi(\mathfrak{p})$ in the lock-ring. If the surface $\mathfrak{F}$ is closed, this description determines the function u uniquely to within an additive constant; for there are no functions regular harmonic at every point of $\mathfrak{F}$ except the constants. Elimination of the arbitrary constant by the normalizing condition that the integral of u^* along the boundary of the hole vanish, is of no further significance for us. But if the nonclosed surface is to be included in our considerations, then we still must prove that the admissible function u actually furnishes the minimum value d of the Dirichlet integral.

Let K_1 be a z-disc concentric with K and with smaller radius. I claim that the limit equation

$$\lim_v D_{K_1}(v - u) = 0 \tag{14.10}$$

holds. On the one hand,

$$\lim_v D_{K_1}(v - \bar{v}) = 0$$

follows from (14.7); on the other hand, the derivatives $\partial\bar{v}/\partial x$ and $\partial\bar{v}/\partial y$ converge uniformly in K_1 to $\partial u/\partial x$ and $\partial u/\partial y$, and hence

$$\lim_v D_{K_1}(\bar{v} - u) = 0.$$

These two limit relations together yield (14.10).

Now associate with each point $\mathfrak{p}$ a local parameter z and a z-disc $K = K(\mathfrak{p}): |z| \leq a$. Also let $B(\mathfrak{p}): |x| \leq b, |y| \leq b$, be a square contained in the interior of $K(\mathfrak{p})$ ($b\sqrt{2} < a$). With the positive number $\theta < 1$ we introduce the block $B^\theta(\mathfrak{p})$ obtained by shrinking $B(\mathfrak{p})$.[12] We can choose countably

---

[11]) Our proof is closely related to that given by Zaremba in his paper cited above.
[12]) The "shrunken block" terminology was introduced in §10: $B^\theta(\mathfrak{p}): |x| \leq \theta b$, $|y| \leq \theta b$.

many points $\mathfrak{p}_i$ such that the interiors of the associated shrunken blocks $B_i^\theta = B^\theta(\mathfrak{p}_i)$ cover $\mathfrak{F}$. Then we determine associated Dieudonné factors μ_i. Carrying out the considerations above for $K = K(\mathfrak{p}_i)$, we find

$$\lim_{v} \int_{\mathfrak{F}} \mu_i \{\text{grad}(v-u)\}^2 = 0. \tag{14.11}$$

While the meaning of this integral is clear, since $v - u$ is continuously differentiable on all of $\mathfrak{F}$, the integral $D_i(v) = \int_{\mathfrak{F}} \mu_i(\text{grad } v)^2$ must be defined. Naturally we mean by this the sum of the integral over all $\mathfrak{F}$ of $\mu_i \Psi(v)$ and the integral (to be computed in the polar coordinates ρ, ψ associated with ζ) of $\mu_i \Psi'(v)$ over the lock-ring. Here $\Psi(v)$ and $\Psi'(v)$ are given by (13.3) and (13.4) (Since the lock-ring $a_0 \leq \rho \leq b_0$ is compact, the second integral actually contributes only for finitely many i.) Because of

$$|\sqrt{D_i(u)} - \sqrt{D_i(v)}| \leq \sqrt{D_i(u-v)},$$

it follows from (14.11) that

$$\lim_{v} \sum_{i=1}^{n} D_i(v) = \sum_{i=1}^{n} D_i(u).$$

The sum $\sum_{i=1}^{n} D_i(v)$ is $\leq D(v)$, and hence we obtain the inequality

$$\sum_{i=1}^{n} D_i(u) \leq d,$$

and, letting $n \to \infty$, we obtain

$$D(u) \leq d.$$

Since $D(u)$ cannot be smaller than d, $D(u) = d$ and $D(u) \leq D(v)$ for every competing function v.

If w is a continuously differentiable function on $\mathfrak{F}$, with a finite Dirichlet integral, then $u + \varepsilon w$ is a competing function for every constant ε. The resulting inequality,

$$D(u+\varepsilon w) \geq D(u) \qquad \text{or} \qquad 2\varepsilon D(u,w) + \varepsilon^2 D(w) \geq 0$$

implies, in the familiar fashion, that the "first variation" $D(u, w)$ vanishes. If we choose $w = v - u$, then the inequality $D(v) \geq D(u)$ is sharpened to

$$D(v) = D(u) + D(v-u).$$

The function U, which $= u$ on the punched surface and $= u^* + \Phi$ on the lid, is a harmonic function that is regular everywhere except at the origin, where it possesses the singularity $\xi/(\xi^2 + \eta^2)$ characteristic of a doublet source. If w is a continuously differentiable function on $\mathfrak{F}$, with a finite Dirichlet integral, and if w vanishes in some neighborhood of $\mathfrak{o}$, then one can form

$$D(U,w) = \int_{\mathfrak{F}} (\operatorname{grad} U, \operatorname{grad} w).$$

I claim that

$$D(U,w) = 0. \tag{14.12}$$

In fact, the equation $D(u, w) = 0$ can be written in the form

$$D_\lambda(uw) + D'_\lambda(uw) = 0.$$

Here

$$\begin{aligned} D_\lambda(uw) &= \int_{\mathfrak{F}} ((1-\lambda)\operatorname{grad} u + \lambda \operatorname{grad} u^*, \operatorname{grad} w) \\ &= D(U,w) - \int_{K'_0} \lambda (\operatorname{grad} \Phi, \operatorname{grad} w), \end{aligned}$$

while $D'_\lambda(u, w)$ is equal to the integral of $\lambda(\operatorname{grad} \Phi, \operatorname{grad} w)$ over the lock-ring. Since the sum is zero, we get for $D(U, w)$ the integral of

$$\lambda(\operatorname{grad} \Phi, \operatorname{grad} w) = (\operatorname{grad} \Phi, \operatorname{grad} w)$$

over the hole K_0. Always bearing in mind that w vanishes in a neighborhood at $\mathfrak{o}$, one uses Green's formula to replace this integral by the integral

$$\left\{2\pi\rho \int_0^1 w \frac{\partial \Phi}{\partial \rho} d\psi \right\}_{\varrho = a_0}$$

over the boundary of the hole. But this vanishes, since $(\partial \Phi/\partial \rho)_{\rho = a_0} = 0$.

§ 15. The elementary differentials

We now restrict ourselves to *closed* Riemann surfaces. By the formula (9.2), a harmonic function U gives rise to an analytic differential $d\tau$,

$$\frac{d\tau}{dz} = \frac{\partial U}{\partial x} - i \frac{\partial U}{\partial y} \qquad (z = x + iy), \tag{15.1}$$

which is regular wherever U is regular. For a given point $\mathfrak{o}$ and an associated local parameter z (called ζ earlier), we have constructed a uniform harmonic function U on the closed Riemann surface $\mathfrak{F}$; U is regular everywhere except at the point $\mathfrak{o}$, and in the neighborhood of $\mathfrak{o}$ it has the form $\mathfrak{R}(1/z)$ + regular harmonic function. Since a harmonic function which is regular *everywhere* on $\mathfrak{F}$ is a constant, this specification determines U to within an additive constant. The analytic differential $d\tau$ arising from U has a singularity only at $\mathfrak{o}$; in a neighborhood of $\mathfrak{o}$ it differs from $-dz/z^2$ by a differential which is regular at $\mathfrak{o}$. The real part $\mathfrak{R}(d\tau)$ is the exact differential dU. Therefore the *period* $\int_\gamma d\tau$ corresponding to an arbitrary closed curve γ (not passing through $\mathfrak{o}$) is *purely imaginary*. This normalization of the periods by separation of the real and imaginary parts is of decisive importance for our investigations. The differential $d\tau$ is uniquely determined by its singularity at $\mathfrak{o}$ plus the requirement that its periods be purely imaginary. Since $d\tau$ has the residue 0 at $\mathfrak{o}$, τ itself is uniform on the class surface $\hat{\mathfrak{F}}$ and regular analytic, except at the points over $\mathfrak{o}$; in the neighborhood of these points it has the form $(1/z)$ + reg. function (z).

U is the potential of the flow on $\mathfrak{F}$ which is generated by a doublet source at $\mathfrak{o}$ of moment 2π in the direction of the positive x-axis. In the same way we can form the potential U' of the flow generated by a doublet source at $\mathfrak{o}$, of the same moment, in the direction of the positive y-axis. For Φ one chooses the function

$$\frac{y}{x^2+y^2}+\frac{y}{a_0^2}$$

in place of (13.1). As the differential $d\tau$ arises from U, so the differential $d\tau'$ arises from U'. The generalization to sources of higher multiplicity is evident. For any one of the natural numbers $n = 1, 2, 3, \ldots$, we form the functions

$$\mathfrak{R}\, z^{-n} = \frac{\mathrm{co}(n\phi)}{r^n}, \qquad -\mathfrak{J}\, z^{-n} = \frac{\mathrm{si}(n\phi)}{r^n}$$

and construct uniform harmonic functions U_n, U_n' which are regular except at $\mathfrak{o}$, at which point their singularity is prescribed by one or the other of the above expressions. To carry out these constructions, we only need to use for Φ one or the other of the functions

$$\frac{\overset{\mathrm{co}}{\underset{\mathrm{si}}{\cdot}}\, n\phi}{r^n}+\frac{r^n\,\overset{\mathrm{co}}{\underset{\mathrm{si}}{\cdot}}\, n\phi}{a_0^{2n}}.$$

We call the functions U_n and U_n', associated with an arbitrary point $\mathfrak{o}$, the *elementary harmonic functions of the second kind*; the associated differentials $d\tau_n$ and $d\tau'_n$ are the *elementary Abelian differentials of the second kind.* The periods of the last are purely imaginary. U_n and U'_n are determined uniquely to within an additive constant by their singularities. Therefore, once the point $\mathfrak{o}$ and the attached local parameter are chosen, $d\tau_n$ and $d\tau'_n$ are determined, and there is no freedom of choice left. Also τ_n and τ'_n are analytic functions on the class surface and they have poles only at the points over $\mathfrak{o}$; the "principal part" at each such pole is z^{-n} or iz^{-n}.

We pass from the flow arising from a multiple source to one coming from *two sources which have the equal but opposite strengths* ± 1. For the sake of simplicity we set the radius a_0 of the hole $= 1$, and assume to begin with that the source and the sink are at two points z_1 and z_2 ($z_1 \neq z_2$) of the hole associated with the local parameter z. We want to construct a harmonic function on $\mathfrak{F}$ which is regular everywhere except at *1* and *2* and which behaves like

$$\frac{1}{2\pi}\log\left|\frac{z-z_2}{z-z_1}\right| = \frac{1}{2\pi}\Re\{\log(z-z_2)-\log(z-z_1)\}$$

at these points.

For this purpose we choose for Φ in the lid that uniform harmonic function which is the real part of

$$\frac{1}{2\pi}\log\frac{(z-z_2)(1-\bar{z}_2 z)}{(z-z_1)(1-\bar{z}_1 z)} = \frac{1}{2\pi}\left\{\log\frac{z-z_2}{z-z_1} + \log\frac{1-\bar{z}_2 z}{1-\bar{z}_1 z}\right\}. \tag{15.2}$$

On the boundary of the hole, where $1/z = \bar{z}$, the function behind the log has positive real values, for there it is

$$= \frac{(z-z_2)(\bar{z}-\bar{z}_2)}{(z-z_1)(\bar{z}-\bar{z}_1)}.$$

Therefore the imaginary part of (15.2) is constant ($= 0$) on the boundary of the hole, and hence the normal derivative of the real part Φ vanishes there. If $\max(|z_1|, |z_2|) = \theta$, $0 < \theta < 1$, then the term

$$\frac{1}{2\pi}\Re\log\frac{1-\bar{z}_2 z}{1-\bar{z}_1 z},$$

from which Φ differs by

$$\frac{1}{2\pi}\Re\log\frac{z-z_2}{z-z_1},$$

is regular in the whole lid, provided that the radius of the lid is taken to be between 1 and θ^{-1}. With this choice of Φ our method of construction for the Dirichlet principle produces a harmonic function U_{12} of the desired type.

Now let $\mathfrak{a}$ and $\mathfrak{a}'$ be any two points on the Riemann surface and let α be a path joining $\mathfrak{a}$ to $\mathfrak{a}'$. By interpolating division points $1 = \mathfrak{a}, 2, 3, \ldots, n = \mathfrak{a}'$ we can split α into arcs $\alpha_{12}, \alpha_{23}, \ldots$, each one of which lies in some z-disc; here z is a local parameter at some point of the arc in question. Consider the sum

$$U_\alpha = U_{\mathfrak{a}\mathfrak{a}'} = U_{12} + U_{23} + \cdots + U_{n-1,n}.$$

If the arc α_{12} lies in a z-disc and the arc α_{23} lies in a z'-disc, then $z - z_2$ and $z' - z_2'$ are local parameters at the point 2. Since $\log\{(z' - z_2')/(z - z_2)\}$ is regular at 2, the two singularities of U_{12} and U_{23} at 2, a source and a sink of the same strength, cancel out in the sum $U_{12} + U_{23}$. The resulting $U_{\mathfrak{a}\mathfrak{a}'}$ is thus a uniform potential function which is regular everywhere except at the two points $\mathfrak{a}$ and $\mathfrak{a}'$; at those points it has sources of equal but opposite strength ± 1. This determines $U_{\mathfrak{a}\mathfrak{a}'}$ uniquely to within an additive constant; not even the choice of the local parameters z and z' at $\mathfrak{a}$ and $\mathfrak{a}'$ has any influence on the result. Again $U_{\mathfrak{a}\mathfrak{a}'}$ gives rise to an analytic differential $d\omega_{\mathfrak{a}\mathfrak{a}'}$ which is regular everywhere except at $\mathfrak{a}$ and $\mathfrak{a}'$; at those points it behaves like $-dz/(2\pi z)$ or $dz'/(2\pi z')$. Of course it would be false to conclude that $d\omega_{\mathfrak{a}\mathfrak{a}'}$ is uniform on the class surface $\mathfrak{F}$; for a small circle κ about the point $\mathfrak{a}$, $\int_\kappa d\omega_{\mathfrak{a}\mathfrak{a}'} =$ the residue $-i$ of $d\omega_{\mathfrak{a}\mathfrak{a}'}$ at $\mathfrak{a}$.

Along with U we considered the potential function U'; now we would like to pair with U_{12} a harmonic function U'_{12} which is singular at the points 1 and 2 only, like

$$\frac{1}{2\pi}\Im\log\frac{z - z_2}{z - z_1} = \phi_2 - \phi_1. \tag{15.3}$$

Here ϕ_1 and ϕ_2 are the azimuths of $z - z_1$ and $z - z_2$. But this function (15.3) is not uniform in the hole; the hole is the unit disc in the z-plane, and z_1 and z_2 are two interior points thereof. Nevertheless, the construction succeeds if we use for Φ the regular harmonic function

$$\Phi = \frac{1}{2\pi}\Im\log\left\{\frac{z - z_2}{z - z_1} : \frac{1 - \bar{z}_2 z}{1 - \bar{z}_1 z}\right\}, \tag{15.4}$$

which is a *uniform* regular harmonic function in the *lock-ring*. Since the two quantities

$$\frac{z - z_1}{1 - \bar{z}_1 z}, \qquad \frac{z - z_2}{1 - \bar{z}_2 z}$$

have absolute value 1 on the boundary $|z| = 1$ of the hole, the real part of the logarithm in (15.4) is constant ($= 0$) on the boundary of the hole, and therefore the imaginary part Φ has the normal derivative 0. Hence our method of construction provides a regular harmonic function u on the punched surface and a regular harmonic function u^* on the lid, which are related by $u = u^* + \Phi$ in the lock-ring. The differential $dU' = dU'_{12}$, which is $= du$ on the punched surface and $= du^* + d\Phi$ on the lid, is a uniform harmonic differential on all of $\mathfrak{F}$; it is regular everywhere except at the points *1* and *2*, where it behaves like $d\phi_2 - d\phi_1$. By means of an appropriate standard subdivision of the curve α into subarcs $\alpha_{12}, \alpha_{23}, \dots, \alpha_{n-1,n}$ one obtains the harmonic differential

$$dU'_\alpha = dU'_{12} + dU'_{23} + \cdots + dU'_{n-1,n}$$

associated with α; it is free of singularities except at $\mathfrak{a}$ and $\mathfrak{a}'$, where it has singularities characteristic of vortices of strength $\pm$ 1. This dU'_α is a harmonic realization of that differential which was denoted by ds_α in § 11. Unlike the treatment there, where we were dealing with arbitrary smooth surfaces, it is unnecessary here to use smoothing processes at the soldered joints *2*, *3*, ..., n – *1*! For a closed curve γ not passing through $\mathfrak{a}$ or $\mathfrak{a}'$, $\int_\gamma dU'_\alpha$ is equal to the intersection number $\mathrm{ch}(\gamma, \alpha)$ of γ with α. We know that in the sense of cohomology, dU'_α does not depend on the construction aids associated with α, nor does it change when α is replaced by another curve α' which joins the same two points $\mathfrak{a}$ and $\mathfrak{a}'$ and which satisfies $\alpha' \sim \alpha$, that is, $\alpha' - \alpha \sim 0$. But *since, in the domain of harmonic differentials, cohomology implies equality,* dU'_α *is uniquely determined by* α. For an arbitrary closed differential dF on $\mathfrak{F}$ we have the formula

$$\int_{\mathfrak{F}} [dU'_\alpha, dF] = \int_\alpha dF. \tag{15.5}$$

dU'_α is the real part of an analytic differential $d\omega'_\alpha$ which is regular everywhere except at $\mathfrak{a}$ and $\mathfrak{a}'$; there it behaves like $-dz/(2\pi iz)$ or $dz'/(2\pi iz')$. For a closed curve γ which does not pass through $\mathfrak{a}$ or $\mathfrak{a}'$, the real part of the period $\int_\gamma d\omega'_\alpha$ is an integer, namely, the intersection number $\mathrm{ch}(\gamma, \alpha)$.

$d\omega_\alpha = d\omega_{\mathfrak{a}\mathfrak{a}'}$ and $d\omega'_\alpha$ are called the *elementary Abelian differentials of the third kind* associated with the path α. The first depends only on the two points $\mathfrak{a}$ and $\mathfrak{a}'$, but not on the joining path α; it vanishes if α is closed. The behavior of $d\omega'_\alpha$ is different.

For a *closed* curve α, dU'_α is a *harmonic differential, regular everywhere without exception ("harmonic differential of the first kind")*, and $d\omega'_\alpha$ is an everywhere regular analytic differential *(Abelian differential of the first kind)*. In this case of a closed curve we write dw_α in place of $d\omega'_\alpha$. For every closed curve γ,

$$\int_\gamma dU'_\alpha = \Re \int_\gamma dw_\alpha = \mathrm{ch}(\gamma, \alpha).$$

If α and β are homologous closed paths, then $dw_\alpha = dw_\beta$; thus dw_α depends only on the cover transformation $S = S_\alpha$ of the class surface associated with α; hence we may denote it by dw_S.

Let $h = 2p$ be the degree of connectivity of the surface, and let $\gamma_1, \dots, \gamma_h$ be a basis for the closed paths. Set

$$dU'_{\gamma_i} = dU'_i, \qquad dw_{\gamma_i} = dw_i.$$

Then $dU'_1, \dots, dU'_h$ are linearly independent harmonic differentials of the first kind, and every such differential dU' is a linear combination of them, with real coefficients c_i,

$$dU' = c_1 dU'_1 + \cdots + c_h dU'_h.$$

Again equality replaces cohomology. The constants c_i are determined by the relations

$$\int_{\gamma_i} dU' = \sum_j s_{ij} c_j \qquad (i,j = 1, \dots, h),$$

where the s_{ij} are the (integral) coefficients $\int_{\gamma_i} dU'_j$ of the skew-symmetric non-degenerate bilinear form, which we have called the *characteristic form.* Accordingly, the h periods $\int_{\gamma_i} dU'$ may be given arbitrarily; they determine the harmonic differential dU' uniquely.

By introducing the conjugate harmonic differential dV' we may pass to the complex Abelian differential of the first kind, $dw = dU' + i\,dV'$. The h differentials $dw_{\gamma_i} = dw_i$ form a basis for the linear space of all Abelian differentials of the first kind, in the sense that every such differential may be expressed in just one way as a linear combination

$$dw = c_1\, dw_1 + \cdots + c_h\, dw_h$$

with real coefficients[13] c_i.

---

[13]) It would have been quite possible, and might have even meant a simplification, to consider only Riemann surfaces and harmonic and Abelian differentials. Then h would have been introduced as the number of linearly independent everywhere regular harmonic differentials. Instead, we subsumed Riemann surfaces under smooth oriented surfaces. I make this remark with regard to the tendency in the literature, sometimes explicit, to construct the theory of Riemann surfaces "without topology." For example, see L. Bieberbach, *Über die Einordnung des Hauptsatzes der Uniformisierung in die Weierstrasssche Funktionentheorie,* Math. Ann., 78 (1918) 312–331. I can see neither difficulties nor any particular profit in this.

While the $s_{jk} = \Re \int_{\gamma_j} dw_k$ from an integral skew-symmetric matrix Σ, the imaginary parts of the periods, $t_{jk} = \Im \int_{\gamma_j} dw_k$ form the symmetric ($t_{jk} = t_{kj}$) coefficient matrix T of a *positive definite quadratic form* $\Sigma\, t_{jk}\, x_j\, x_k$. In order to see this, form, for two closed curves α and β, the Dirichlet integral

$$\int_{\mathfrak{F}} (dU'_\alpha, dU'_\beta) = \int_{\mathfrak{F}} [dU'_\alpha, dV'_\beta] = \int_\alpha dV'_\beta = \Im \int_\alpha dw_\beta .$$

This gives the symmetry law

$$\Im \int_\alpha dw_\beta = \Im \int_\beta dw_\alpha \tag{15.6}$$

immediately, and also: if $dU' = \Sigma\, x_j\, dU'_j$ is any harmonic differential of the first kind with arbitrary real coefficients x_j, then the Dirichlet integral $\int (dU', dU')$ has the value $\Sigma\, t_{jk}\, x_j\, x_k$. Hence the quadratic form is positive unless $x_1 = \cdots = x_h = 0$. If one chooses an integral basis for γ_j, then the integral skew-symmetric form $\Sigma\, s_{jk}\, y_j\, x_k$ and the positive definite quadratic form $\Sigma\, t_{jk}\, x_j\, x_k$ are determined up to an arbitrary unimodular transformation of the variables x_j and y_j.

The separation into real and imaginary parts, characteristic for our construction, has the consequence that linear independence is interpreted as linear independence with *real* coefficients. For complex differentials it seems more natural to admit any *complex* coefficients. From the h differentials of the first kind, $dw_1 \ldots, dw_h$, which are linearly independent in the real sense, we may choose a maximal number, $dw_1, \ldots, dw_q$, which are linearly independent in the complex sense. Then

$$dw_1, -i\, dw_1, \ldots, dw_q, -i\, dw_q$$

form a basis in the real sense; hence $h = 2q$; in other words, q is the genus p. The function theorist will be inclined to see in this relation the intrinsic reason why the degree of connectivity h is an even number. Without leaving the "real point of view" one can also argue as follows. The differentials, $i\, dw_j$, obtained by multiplying the h differentials $dw_j = dw_{\gamma_j}$ by the imaginary unit i, must be expressible as linear combinations of the dw_k with *real* coefficients r_{jk}:

$$i\, dw_j = \sum_k r_{jk}\, dw_k \qquad (i, k = 1, \ldots, h). \tag{15.7}$$

Since two successive multiplications by i amount to multiplication by

$i^2 = -1$, the matrix $R = \|r_{jk}\|$ must satisfy the equation $R^2 = -E$, where E is the identity matrix. From this it follows that

$$(\det R)^2 = (-1)^h.$$

Since the left-hand side is positive, we again get the result that h must be even. If one integrates the equations (15.7) over γ_l and takes the real part on each side, then one obtains

$$-\mathfrak{I}\int_{\gamma_l} dw_j = \sum_k r_{jk} \mathfrak{R} \int_{\gamma_l} dw_k$$

or

$$t_{jl} = \sum_k r_{jk} s_{kl}, \qquad T = R\mathbf{\Sigma}.$$

Thus the following algebraic problem is associated with the theory of Abelian differentials of the first kind: to find an integral, nondegenerate, skew-symmetric form $\mathbf{\Sigma}$ and a positive, definite, quadratic form T, with coefficients matrices $\mathbf{\Sigma}$ and T, such that the matrix $R = T\mathbf{\Sigma}^{-1}$ satisfies the equation $R^2 = -E$. Two solutions of this problem, $\mathbf{\Sigma}$, T, and $\mathbf{\Sigma}_1$, T_1, rate as equivalent if $\mathbf{\Sigma}_1 = A\mathbf{\Sigma}A'$ and $T_1 = ATA'$, where A is an arbitrary unimodular matrix (and A' is its transpose).

An analytic differential $d\omega$ which is regular on $\mathfrak{F}$ except for a finite number of poles will be called an *Abelian differential (of the third kind)*. If no nonzero residues arise, then one talks of a *differential of the second kind*; if there are no poles at all, then one of the *first kind*. I prefer this terminology, in which the three kinds are not disjoint, but all differentials of the first kind are also second kind, and the differentials of the second kind are also third kind. At a point $\mathfrak{o}$ with local parameter z one has the development

$$d\omega = (A_{-m} z^{-m} + \cdots + A_{-1} z^{-1} + A_0 + A_1 z + \cdots)\, dz.$$

The residue is $2\pi i\, A_{-1}$, and

$$(A_{-m} z^{-m} + \cdots + A_{-1} z^{-1})\, dz \tag{15.8}$$

is the *principal part* of the differential at $\mathfrak{o}$. The residue is independent of the choice of the local parameter. The sum of the residues at the different poles is necessarily zero. If $A_{-1} = 0$, then one can build an Abelian differential of the second kind that has a pole only at $\mathfrak{o}$ and has there the given principal part (15.8). For use a linear combination, with suitable complex coefficients, of the elementary differentials $d\tau_1, \ldots, d\tau_m$ attached to the point $\mathfrak{o}$. If one is given l points $\mathfrak{p}_1, \ldots, \mathfrak{p}_l$ and l complex numbers $C_1, \ldots, C_l$ whose sum is zero, then choose a fixed point $\mathfrak{o}$ distinct from the $\mathfrak{p}_i$ and form the linear combination

$$C_1 d\omega_{\mathfrak{op}_1} + \cdots + C_l d\omega_{\mathfrak{op}_l};$$

this is a differential which has poles of the first order at $\mathfrak{p}_1, \ldots, \mathfrak{p}_l$ with the given residues $C_1, \ldots, C_l$ (since the sum of the C is zero, this differential is regular at $\mathfrak{o}$). From all this it follows that *for arbitrarily located poles with arbitrarily given principal parts, one can always form an Abelian differential, provided only that the sum of the residues at the given poles is zero.* The differential thus determined is unique to within an additive arbitrary differential of the first kind. We have arranged our investigation from the start so that it allows us a clear view, not only of the linear manifold of differentials, but also of their integrals; the separation into real and imaginary parts becomes essential only for the question of periods.

§ 16. The symmetry laws

For the Abelian integrals of the first kind we found: a law of *skew symmetry,* stating that the intersection number of two closed curves α and β depends skew symmetrically on α and β; and a law of symmetry, expressed in the equation (15.6). The first is valid for arbitrary *oriented smooth surfaces*; the second is specific to Riemann surfaces. These laws are special cases of the laws which hold for the elementary integrals of the third kind. Let α and β be two (closed or) open curves, the first joining $\mathfrak{p}_0$ to $\mathfrak{p}_1$ and the second joining $\mathfrak{q}_0$ to $\mathfrak{q}_1$. It is assumed that α does not pass through $\mathfrak{q}_0$ or $\mathfrak{q}_1$ and that β does not pass through $\mathfrak{p}_0$ or $\mathfrak{p}_1$. As is proper, the definition of the anti-symmetric intersection number of α and β has already been treated in the first topological part. There follow now the laws of symmetry which, in the older literature, were lumped into one, the "law of interchange of argument and parameter."

Obviously, the Dirichlet integral

$$D(U_\alpha, U_\beta) = \int (dU_\alpha, dU_\beta) = \int [dU_\alpha, dV_\beta]$$

depends symmetrically on α and β. Since the differential dV_β, conjugate to dU_β, has the residues -1 and $+1$ at the points $\mathfrak{q}_0$ and $\mathfrak{q}_1$, it follows from (11.11) that

$$\int (dU_\alpha, dU_\beta) = -\int_\beta dU_\alpha = -\Re \int_\beta d\omega_\alpha;$$

hence

$$\Re \int_\alpha d\omega_\beta = \Re \int_\beta d\omega_\alpha$$

or

$$[\mathfrak{R}\,\omega_{\mathfrak{q}_0\mathfrak{q}_1}(\mathfrak{p})]_{\mathfrak{p}=\mathfrak{p}_0}^{\mathfrak{p}_1} = [\mathfrak{R}\,\omega_{\mathfrak{p}_0\mathfrak{p}_1}(\mathfrak{q})]_{\mathfrak{q}=\mathfrak{q}_0}^{\mathfrak{q}_1}.$$

On the left, $\mathfrak{q}_0$ and $\mathfrak{q}_1$ are the parameters and $\mathfrak{p}_0$ and $\mathfrak{p}_1$ are the arguments; their roles are switched on the right.

Secondly, we consider

$$\int (dU_\alpha, dU'_\beta) = \int [dU_\alpha, dV'_\beta].$$

Now dV'_β certainly has singularities at $\mathfrak{q}_0$ and $\mathfrak{q}_1$, but its residues are zero; hence the value of this surface integral is zero. On the other hand,

$$\int (dU_\alpha, dU'_\beta) = \int [dU'_\beta, dV_\alpha],$$

and, by the formula (11.10), this is equal to

$$\{\int_\alpha dU'_\beta - \mathrm{ch}(\alpha, \beta)\} - \int_\beta dV_\alpha.$$

Hence

$$\mathfrak{R}\int_\alpha d\omega'_\beta \equiv \mathfrak{J}\int_\beta d\omega_\alpha \quad (\text{mod } 1)$$

and naturally also

$$\mathfrak{J}\int_\alpha d\omega_\beta \equiv \mathfrak{R}\int_\beta d\omega'_\alpha \quad (\text{mod. } 1)$$

More precisely, each of the differences

$$\mathfrak{R}\int_\alpha d\omega'_\beta - \mathfrak{J}\int_\beta d\omega_\alpha \qquad \text{and} \qquad \mathfrak{J}\int_\alpha d\omega_\beta - \mathfrak{R}\int_\beta d\omega'_\alpha$$

is equal to the intersection number ch(α, β) of the two paths α and β. Finally,

$$\int (dU'_\alpha, dU'_\beta) = \int [dU'_\alpha, dV'_\beta],$$

and, since dV'_β has zero residues, this is equal to the line integral $\int_\alpha dV'_\beta$. Hence

$$\mathfrak{J}\int_\alpha d\omega'_\beta = \mathfrak{J}\int_\beta d\omega'_\alpha.$$

We summarize these laws in one formula:

$$\text{(I)} \qquad \begin{cases} \mathfrak{R}\int_\alpha d\omega_\beta = \mathfrak{R}\int_\beta d\omega_\alpha, \qquad \mathfrak{J}\int_\alpha d\omega'_\beta = \mathfrak{J}\int_\beta d\omega'_\alpha; \\ \mathfrak{R}\int_\alpha d\omega'_\beta - \mathfrak{J}\int_\beta d\omega_\alpha = \mathfrak{J}\int_\alpha d\omega_\beta - \mathfrak{R}\int_\beta d\omega'_\alpha = \mathrm{ch}(\alpha, \beta). \end{cases}$$

If the path β is *closed* $\{d\omega'_\beta = dw_\beta\}$, these equations reduce to

$$(\mathrm{I}_0) \qquad \Im\int_\alpha d\omega_\beta = \Im\int_\beta d\omega'_\alpha, \qquad \Re\int_\alpha d\omega_\beta \equiv \Im\int_\beta d\omega_\alpha \quad (\mathrm{mod}\,1);$$

if α is also closed they reduce to the symmetry relation, derived in the preceding paragraph, for the integrals of the first kind:

$$(\mathrm{I}_{00}) \qquad \Im\int_\alpha d\omega_\beta = \Im\int_\beta d\omega_\alpha .$$

The simplest of the elementary integrals of the second kind are derived from the elementary integral of the third kind by letting the pair of sources $\mathfrak{q}_0$ and $\mathfrak{q}_1$ coalesce into a doublet source. Let $\zeta = \xi + i\eta$ be a local parameter at $\mathfrak{q}_0$, and let $\mathfrak{q}_1$ be the point $\zeta = \varepsilon > 0$; while $\varepsilon \to 0$, the point $\mathfrak{q}_1$ tends to $\mathfrak{q}_0$ along the real axis of the ζ-plane. For the successive derivatives $d/d\zeta$, $d^2/d\zeta^2$, ... at the point $\zeta = 0$ we write, dropping the index 0 on $\mathfrak{q}_0$, $d/d\mathfrak{q}$, $d^2/d\mathfrak{q}^2$, ... The elementary differentials of the second kind associated with the local parameter ζ at $\mathfrak{q} = \mathfrak{q}_0$ may now be denoted more precisely by the notation $d\tau_{\mathfrak{q},n}$ and $d\tau'_{\mathfrak{q},n}$ which specifies the pole. For the construction of the potential function $U_{\mathfrak{q}_0\mathfrak{q}_1}$ we used the unit disc in the ζ-plane for the *hole*, and for the *jump* over the boundary of the hole we used the real part $\Phi_{\mathfrak{q}_0\mathfrak{q}_1} = \Phi_\varepsilon$ of the function

$$(16.1) \qquad \frac{1}{2\pi}\log\frac{(\zeta-\varepsilon)(1-\varepsilon\zeta)}{\zeta}.$$

The development of this function in powers of the parameter ε runs as follows:

$$-\frac{1}{2\pi}\sum_{n=1}^{\infty}\frac{\varepsilon^n}{n}(\zeta^n+\zeta^{-n}).$$

Hence

$$(16.2) \qquad U_{\mathfrak{q}_0\mathfrak{q}_1} = -\frac{1}{2\pi}\sum_{n=1}^{\infty}\frac{\varepsilon^n}{n}\Re\,\tau_{\mathfrak{q},n}.$$

The development of $[U_\alpha(\mathfrak{q})]_{\mathfrak{q}_0}^{\mathfrak{q}_1} = [U_\alpha(\zeta)]_{\zeta=0}^{\zeta=\varepsilon}$ in powers of ε is

$$\sum_{n=1}^{\infty}\frac{\varepsilon^n}{n!}\Re\frac{d^n\omega_\alpha}{d\mathfrak{q}^n}.$$

From the first equation in (I),

$$\int_{\alpha} dU_{\mathfrak{q}_0 \mathfrak{q}_1}(\mathfrak{p}) = \int_{\mathfrak{q}_0}^{\mathfrak{q}_1} dU_{\alpha}(\mathfrak{q})\,;$$

it then follows by comparing the coefficients of ε^n in the power series that

$$\Re \int_{\alpha} d\tau_{\mathfrak{q},n}(\mathfrak{p}) = \frac{-2\pi}{(n-1)!}\,\Re\,\frac{d^n \omega_{\alpha}(\mathfrak{q})}{d\mathfrak{q}^n}. \tag{16.3}$$

Here α is any curve that does not pass through the point $\mathfrak{q}$. Similarly, from the last of the four equations (I),

$$\Im \int_{\alpha} d\tau_{\mathfrak{q},n}(\mathfrak{p}) = \frac{-2\pi}{(n-1)!}\,\Re\,\frac{d^n \omega'_{\alpha}(\mathfrak{q})}{d\mathfrak{q}^n}. \tag{16.4}$$

Let $\mathfrak{p}_0$ be an interior point of the punched surface and let z be a local parameter at $\mathfrak{p}_0$. Let $|z - z_0| < a$ define a z-disc K about $\mathfrak{p}_0$, contained in the punched surface, and let $|z - z_0| \leq \theta a$ be a concentric disc K_θ of smaller radius ($\theta < 1$). To justify our procedure completely we must supplement the expansion (16.2) with an estimate of the remainder by showing that

$$2\pi\,\frac{d\omega_{\mathfrak{q}_0 \mathfrak{q}_1}}{dz} + \sum_{\nu=1}^{n-1} \frac{\varepsilon^\nu}{\nu}\,\frac{d\tau_{\mathfrak{q},\nu}}{dz} \tag{16.5}$$

is in absolute value less than a constant times ε^n in the disc K_θ, where the constant depends neither on ε nor on z. If α is an arbitrary curve, not through $\mathfrak{q}$, then one can choose the hole with center $\mathfrak{q}$ so small that α lies in the punched surface; it follows then, from such an estimate based on a suitable standard subdivision of α, that

$$2\pi \int_{\alpha} d\omega_{\mathfrak{q}_0 \mathfrak{q}_1} + \sum_{\nu=1}^{n-1} \frac{\varepsilon^\nu}{\nu} \int_{\alpha} d\tau_{\mathfrak{q},\nu} = O(\varepsilon^n).$$

Set

$$\log \frac{(\zeta - \varepsilon)(1 - \varepsilon\zeta)}{\zeta} + \sum_{\nu=1}^{n-1} \frac{\varepsilon^\nu}{\nu}(\zeta^\nu + \zeta^{-\nu}) = F(\zeta), \qquad \Phi = \Re F.$$

The function u, which

$$= U = 2\pi\, U_{\mathfrak{q}_0 \mathfrak{q}_1} + \sum_{\nu=1}^{n-1} \frac{\varepsilon^\nu}{\nu}\, U_{\mathfrak{q},\nu}$$

in the punched surface, and $= U - \Phi$ in the hole, is then that one of the functions with jump Φ over the boundary of the hole which minimizes the Dirichlet integral. One uses, as a comparison function, the function v which vanishes outside the hole and which, in the hole, is the regular harmonic function with boundary values $-\Phi$ on the boundary of the hole. (For a justification of this procedure see page 108.) This function is $= -\Re G(\zeta)$, where

$$\tfrac{1}{2} G(\zeta) = \log(1 - \varepsilon\zeta) + \sum_{\nu=1}^{n-1} \frac{\varepsilon^\nu}{\nu} \zeta^\nu.$$

Its Dirichlet integral is equal to the integral of $| dG/d\zeta |^2$ over the unit disc. Since

$$\frac{dG}{d\zeta} = -2 \frac{\varepsilon^n \zeta^{n-1}}{1 - \varepsilon\zeta},$$

we get for this integral the value

$$8\pi\varepsilon^{2n} \int_0^1 \int_0^1 \frac{\rho^{2n-1}\, d\rho\, d\phi}{1 + (\varepsilon\rho)^2 - 2\varepsilon\rho \operatorname{co}\phi} = 8\pi\varepsilon^{2n} \int_0^1 \frac{\rho^{2n-1}\, d\rho}{1 - \varepsilon^2\rho^2} \leq \varepsilon^{2n} \frac{4\pi}{n} \frac{1}{1 - \varepsilon^2}.$$

Thus the Dirichlet integral of u over the disc K is smaller than this value, and hence one obtains, according to (12.15), the upper bound

$$\frac{\varepsilon^{2n}}{na^2} \frac{4}{(1 - \varepsilon^2)(1 - \theta^2)}$$

for the square of the absolute value of (16.5). Thus the estimate desired is proved.

Applying the same procedure to the second and third of the equations (I) we extend (16.3) and (16.4) to the complete theorem consisting of the following four relations:

$$(\mathrm{II}^n) \quad \left\{ \begin{aligned} \Re \int_\alpha d\tau_{\mathfrak{q},n}(\mathfrak{p}) &= \frac{-2\pi}{(n-1)!} \Re \frac{d^n \omega_\alpha(\mathfrak{q})}{d\mathfrak{q}^n}, \\ \Im \int_\alpha d\tau_{\mathfrak{q},n}(\mathfrak{p}) &= \frac{-2\pi}{(n-1)!} \Re \frac{d^n \omega'_\alpha(\mathfrak{q})}{d\mathfrak{q}^n}, \\ \Re \int_\alpha d\tau'_{\mathfrak{q},n}(\mathfrak{p}) &= \frac{-2\pi}{(n-1)!} \Im \frac{d^n \omega_\alpha(\mathfrak{q})}{d\mathfrak{q}^n}, \\ \Im \int_\alpha d\tau'_{\mathfrak{q},n}(\mathfrak{p}) &= \frac{-2\pi}{(n-1)!} \Im \frac{d^n \omega'_\alpha(\mathfrak{q})}{d\mathfrak{q}^n}. \end{aligned} \right.$$

For a *closed* curve α, the second and fourth of these relations give the equations

$$(\mathrm{II}_0^n) \quad \begin{cases} \int_\alpha d\tau_{\mathfrak{q},n}(\mathfrak{p}) = \dfrac{-2\pi i}{(n-1)!}\,\mathfrak{R}\dfrac{d^n w_\alpha(\mathfrak{q})}{d\mathfrak{q}^n}, \\ \int_\alpha d\tau'_{\mathfrak{q},n}(\mathfrak{p}) = \dfrac{-2\pi i}{(n-1)!}\,\mathfrak{J}\dfrac{d^n w_\alpha(\mathfrak{q})}{d\mathfrak{q}^n}. \end{cases}$$

Thus the periods of the elementary differentials of the second kind, $d\tau_n$ and $d\tau'_n$, on the closed path α, are simply expressed by the real and imaginary parts of the nth derivative of the integral of the first kind, w_α, at the pole.

Let $\mathfrak{p}_0 = \mathfrak{p}$ be a point $\neq \mathfrak{q}$, let z be a local parameter at $\mathfrak{p}$, and interpret the symbol $d^n/d\mathfrak{p}^n$ as d^n/dz^n at $z = 0$. If we take α to be the segment from $\mathfrak{p}_0$, $z = 0$, to $\mathfrak{p}_1$, $z = \varepsilon > 0$, in the z-disc, then the equations (II^n), developed in powers of ε, give[14]

$$(\mathrm{III}^{m,n}) \quad \begin{cases} \dfrac{1}{(m-1)!}\,\mathfrak{R}\dfrac{d^m\tau_{\mathfrak{q},n}(\mathfrak{p})}{d\mathfrak{p}^m} = \dfrac{1}{(n-1)!}\,\mathfrak{R}\dfrac{d^n\tau_{\mathfrak{p},m}(\mathfrak{q})}{d\mathfrak{q}^n}, \\ \dfrac{1}{(m-1)!}\,\mathfrak{J}\dfrac{d^m\tau_{\mathfrak{q},n}(\mathfrak{p})}{d\mathfrak{p}^m} = \dfrac{1}{(n-1)!}\,\mathfrak{R}\dfrac{d^n\tau'_{\mathfrak{p},m}(\mathfrak{q})}{d\mathfrak{q}^n}, \\ \dfrac{1}{(m-1)!}\,\mathfrak{R}\dfrac{d^m\tau'_{\mathfrak{q},n}(\mathfrak{p})}{d\mathfrak{p}^m} = \dfrac{1}{(n-1)!}\,\mathfrak{J}\dfrac{d^n\tau_{\mathfrak{p},m}(\mathfrak{q})}{d\mathfrak{q}^n}, \\ \dfrac{1}{(m-1)!}\,\mathfrak{J}\dfrac{d^m\tau'_{\mathfrak{q},n}(\mathfrak{p})}{d\mathfrak{p}^m} = \dfrac{1}{(n-1)!}\,\mathfrak{J}\dfrac{d^n\tau'_{\mathfrak{p},m}(\mathfrak{q})}{d\mathfrak{q}^n} \end{cases} \quad (m, n = 1, 2, \ldots).$$

§ 17. The uniform functions on $\mathfrak{F}$ as a subspace of the additive and multiplicative functions on $\hat{\mathfrak{F}}$. The Riemann-Roch theorem

By a *transcendental normed Abelian differential of the second kind, dv,* we mean such an analytic differential of the second kind with purely imaginary periods. Associated with dv there is a function $v(\hat{\mathfrak{p}})$, regular analytic except for poles on the class surface $\hat{\mathfrak{F}}$ *(additive function* or *transcendental normed Abelian integral of the second kind),* which behaves additively relative to the cover transformations S of $\hat{\mathfrak{F}}$:

$$v(\hat{\mathfrak{p}}\,S) = v(\hat{\mathfrak{p}}) + \pi_s,$$

[14]) Earlier it was customary to derive these symmetry laws (following Riemann and C. Neumann) by integrating around the boundary of the canonically cut surface. Our normalization of the elementary integrals renders this dissection superfluous. I regard this as an essential improvement of the method.

with pure imaginary constant periods π_S. For any two cover transformations, S and T, the periods are related by the rule $\pi_{ST} = \pi_S + \pi_T$.

There are only finitely many points $\mathfrak{q}$ on $\mathfrak{F}$ over which an additive function v has poles; if z is a local parameter at $\mathfrak{q}$, and hence at all points of $\hat{\mathfrak{F}}$ over $\mathfrak{q}$, then v has the same principal part at all of these points over $\mathfrak{q}$:

$$\frac{A_{-1}}{z} + \frac{A_{-2}}{z^2} + \cdots + \frac{A_{-r}}{z^r}. \tag{17.1}$$

By the formulae (II_0^n) the periods π_S of $\tau_{\mathfrak{q},n}$ and $\tau'_{\mathfrak{q},n}$ are given by

$$\frac{-2\pi i}{(n-1)!}\,\Re\,\frac{d^n w_S(\mathfrak{q})}{d\mathfrak{q}^n} \quad \text{and} \quad \frac{-2\pi i}{(n-1)!}\,\Im\,\frac{d^n w_S(\mathfrak{q})}{d\mathfrak{q}^n}. \tag{17.2}$$

If a_k, a'_k $(k = 1, \ldots, r)$ are arbitrary real constants, then

$$v_{\mathfrak{q}} = \sum_{k=1}^{r} (a_k \tau_{\mathfrak{q},k} - a'_k \tau'_{\mathfrak{q},k}) \tag{17.3}$$

is an additive function which has a pole only at $\mathfrak{q}$; if we set $A_{-k} = a_k + ia'_k$, then it has the principal part (17.1). Thus the poles and the principal parts of an additive function may be specified arbitrarily; they determine the function uniquely to within an arbitrary additive constant. The function is represented as the sum of the expressions (17.3) corresponding to the different poles $\mathfrak{q}$:

$$v = \textstyle\sum_{\mathfrak{q}} v_{\mathfrak{q}} + \text{const}. \tag{17.4}$$

This representation is the analog of the *partial fraction decomposition* of a rational function. Thus to obtain this analog we must extend the class of meromorphic functions on the surface $\mathfrak{F}$ to the larger class of "additive functions."

Besides the partial fraction decomposition, the representation of a rational function as a *product of linear factors* is important. To extend this from the sphere to arbitrary closed Riemann surfaces we must discuss "*multiplicative functions*"; that is, those functions Θ which are uniform and regular except for poles on the covering surface $\hat{\mathfrak{F}}$, and which behave multiplicatively relative to any cover transformation S of $\hat{\mathfrak{F}}$:

$$\Theta(\hat{\mathfrak{p}}\,S) = \mu_S \Theta(\hat{\mathfrak{p}}),$$

where the multiplier μ_S is a constant, of absolute value one, associated with

S.[15] The μ_S form a system of characters; that is, for any two cover transformations S and T of $\mathfrak{F}$, we have $\mu_{ST} = \mu_S \mu_T$. At all points over a given point $\mathfrak{p}$, of $\mathfrak{F}$, Θ has the same order; this may be positive (then Θ has a zero over $\mathfrak{p}$) or negative (Θ has a pole over $\mathfrak{p}$) or zero. This order is nonzero only at finitely many points $\mathfrak{p}$ of $\mathfrak{F}$. A multiplicative function without zeros and poles is necessarily a constant; for then $\log |\Theta| = \Re \log \Theta$ is a uniform regular harmonic function on $\mathfrak{F}$. Thus a multiplicative function is uniquely determined, to within an arbitrary nonzero constant factor, by its zeros and poles. The number of zeros is equal to the number of poles, if each zero and pole is counted with the right multiplicity. For $d\Theta/\Theta$ is an Abelian differential of the third kind on $\mathfrak{F}$, and the sum of its residues is = the number of zeros minus the number of poles of Θ. Finally, a finite number of points of $\mathfrak{F}$ over which the poles and zeros of Θ lie, and their multiplicities, may be specified arbitrarily, subject to the one condition that the total multiplicity of the zeros must agree with the total multiplicity of the poles. To show this we construct the *elementary multiplicative function* $\Theta_{\mathfrak{a}\mathfrak{b}}$ [the analog of the function $(z-a)/(z-b)$ on the z-sphere] which has a zero of order 1 over the point $\mathfrak{a}$, a pole of order 1 over the point $\mathfrak{b}$, and is otherwise regular and nonzero. We set

$$\Theta_{\mathfrak{a}\mathfrak{b}} = e^{-2\pi\omega_{\mathfrak{a}\mathfrak{b}}}. \tag{17.5}$$

In fact, for an arbitrary closed curve γ on $\mathfrak{F}$, not through $\mathfrak{a}$ or $\mathfrak{b}$, we have from (I_0) the congruence

$$-2\pi \int_\gamma d\omega_{\mathfrak{a}\mathfrak{b}} \equiv -2\pi i \int_{\mathfrak{a}}^{\mathfrak{b}} dw_\gamma \qquad (\mathrm{mod}\, 2\pi i). \tag{17.6}$$

If $\gamma \sim 0$, then $dw_\gamma = 0$, and by (17.6)

$$e^{-2\pi \int_\gamma d\omega_{\mathfrak{a}\mathfrak{b}}} = 1.$$

That is, (17.5) is a *uniform* function on $\hat{\mathfrak{F}}$. If the arbitrary closed path γ (not through $\mathfrak{a}$ and $\mathfrak{b}$) corresponds to the cover transformation $S = S_\gamma$ of $\hat{\mathfrak{F}}$, then the equation (17.6) shows that the multiplier μ_S of $\Theta_{\mathfrak{a}\mathfrak{b}}$ has the value

$$e^{2\pi i \Re \int_{\mathfrak{b}}^{\mathfrak{a}} dw_S} = \mathrm{ex}\left(\Re \int_{\mathfrak{b}}^{\mathfrak{a}} dw_S\right).$$

---

[15]) See Riemann, *Theorie der Abelschen Funktionen*, Werke (2nd edition), p. 140, and especially Appell, Journal de mathématiques, 3. ser., **9** (1883), and Acta Mathematica, **13** (1890).

If $\mathfrak{p}_1, \mathfrak{p}_2, \dots, \mathfrak{p}_r$ are the prescribed zeros and $\mathfrak{q}_1, \dots, \mathfrak{q}_r$ the prescribed poles of a multiplicative function $\boldsymbol{\Theta}$ (with each zero and pole entered as many times as its multiplicity demands), then one obtains the sought for $\boldsymbol{\Theta}$ from the equation

$$\boldsymbol{\Theta} = \text{Const}\, \boldsymbol{\Theta}_{\mathfrak{p}_1 \mathfrak{q}_1} \cdot \boldsymbol{\Theta}_{\mathfrak{p}_2 \mathfrak{q}_2} \cdots \boldsymbol{\Theta}_{\mathfrak{p}_r \mathfrak{q}_r}; \tag{17.7}$$

this equation is the analog of the product representation of a rational function. We also write this equation in the symbolic form

$$\boldsymbol{\Theta} = \frac{\mathfrak{p}_1 \mathfrak{p}_2 \cdots \mathfrak{p}_r}{\mathfrak{q}_1 \mathfrak{q}_2 \cdots \mathfrak{q}_r}.$$

Naturally, we have tacitly assumed that all the $\mathfrak{p}_i$ were different from all the $\mathfrak{q}_i$. If this is not the case, then the symbol on the right-hand side is to be given that meaning obtained by "cancelling" (as with common fractions) symbols common to the numerator and denominator.

The theorem that a multiplicative function vanishes as many times as it becomes infinite applies in particular to every function f which is uniform and without essential singularities on $\mathfrak{F}$. If we apply the theorem to $f - a$, where a is an arbitrary constant, we see that *a function uniform and regular except for poles on $\mathfrak{F}$ assumes each value, including ∞, equally often.*

In the case $p = 0$, the integral of the second kind, $\tau_{\mathfrak{q}}$, on $\mathfrak{F}$ is itself uniform; it has a single pole, of order one; hence it assumes each value exactly once, and maps $\mathfrak{F}$ conformally onto the sphere. *The sphere is the only Riemann surface of genus zero.*

Now we want to investigate in general *how the meromorphic functions on $\mathfrak{F}$* (that is, the uniform functions regular except for poles on $\mathfrak{F}$) *fit in as special cases among the additive and multiplicative functions;* we shall do this by setting up the conditions for the representations (17.4) and (17.7) to give uniform functions on $\mathfrak{F}$. This will lead us to the *Riemann-Roch theorem* and *Abel's theorem*, respectively. We begin with (17.4) and consider first, for the sake of simplicity, only *simple* poles. So let $\mathfrak{p}_1, \mathfrak{p}_2, \dots, \mathfrak{p}_m$ be any m distinct points on $\mathfrak{F}$. Uniform functions on $\mathfrak{F}$ which have at worst poles of order one at the points $\mathfrak{p}_1, \dots, \mathfrak{p}_m$ and are otherwise regular, must be of the form

$$f = (a_1 \tau_{\mathfrak{p}_1} + a_1' \tau_{\mathfrak{p}_1}') + (a_2 \tau_{\mathfrak{p}_2} + a_2' \tau_{\mathfrak{p}_2}') + \cdots + (a_m \tau_{\mathfrak{p}_m} + a_m' \tau_{\mathfrak{p}_m}') + (b + ib'),$$

where a_i, a_i', b, b' are real constants. The condition that the $2p$ periods of f, corresponding to a basis of $2p$ closed paths γ_l, must vanish, leads to $2p$ *homogeneous linear equations for the a_i, a_i'* with real coefficients. These equations are certainly satisfied by some constants a_i, a'_i, not all zero, if $m > p$.

If one specifies arbitrarily more than p points on $\mathfrak{F}$*, then there exist nonconstant uniform functions on* $\mathfrak{F}$ *which have at most poles of order one at these points and are otherwise regular. In particular, this demonstrates the existence of nonconstant uniform functions, regular except for poles, on any closed Riemann surface.*

If $\mathfrak{p}_1, \mathfrak{p}_2, \ldots, \mathfrak{p}_r$ are any r distinct points on $\mathfrak{F}$ and if $m_1, m_2, \ldots, m_r$ are associated integers (> 0, $= 0$, or < 0), then any function or differential on $\mathfrak{F}$ which is of order at least m_i at the point $\mathfrak{p}_i$ and is otherwise regular is called a *multiple of the divisor*[16]

$$\mathfrak{d} = \mathfrak{p}_1^{m_1} \mathfrak{p}_2^{m_2} \cdots \mathfrak{p}_r^{m_r}. \tag{17.8}$$

A divisor is a symbol made up of a finite number of points and associated order numbers (exponents). It is useful to make the agreement that we multiply and divide such divisors in the same way as ordinary fractions. The sum of the exponents,

$$\sum_{i=1}^{r} m_i = m$$

is the *(total) order* of the divisor. The theorem just above may be generalized as follows.

If the order m of a divisor $\mathfrak{d}$ *is not less than p, then there are uniform functions on* $\mathfrak{F}$ *which are multiples of* $1/\mathfrak{d}$; *actually there are at least* $m + 1 - p$ *which are linearly independent* (in the complex sense).

If the first s exponents m_i are positive and the rest negative, then those additive functions which have poles over $\mathfrak{p}_i$ of order at most m_i ($i = 1, 2, \ldots, s$), and are otherwise regular, depend in a homogeneous linear fashion on

$$2(1 + \sum_{i=1}^{s} m_i)$$

real constants. The demand that all the periods of the additive function should

---

[16]) This terminology comes from the arithmetic theory of algebraic functions of Hensel and Landsberg; see Hensel and Landsberg, *Theorie der algebraischen Funktionen einer Variablen*, Leipzig, 1902.

vanish imposes $2p$ linear homogeneous real equations on these constants. The demand that the function vanish with order at least $-m_i$ at the point $\mathfrak{p}_i\,(i = s+1, \dots, r)$ imposes

$$2 \cdot \sum_{i=s+1}^{r} (-m_i)$$

further equations of the same type. The excess of the number of unknowns over the number of equations is

$$2(1 + \sum_{i=1}^{s} m_i) - 2p - 2 \sum_{i=s+1}^{r} (-m_i) = 2(m + 1 - p).$$

This gives a lower bound for the number of solutions which are linearly independent in the real sense.

The most complete answer to the question is contained in the Riemann-Roch theorem.[17] We assume first that the divisor $\mathfrak{d}$ is integral: $m_j \geq 0$. The differentials of the first kind which correspond to a given basis of closed paths, $\gamma_1, \dots, \gamma_h$, will again be denoted by $dw_i = dw_{\gamma_i}$. The most general additive function which is a multiple of $1/\mathfrak{d}$ is given by

$$\sum_{j=1}^{r} \sum_{k=1}^{m_j} \{a_{jk}\tau_{\mathfrak{p}_j,k} + a'_{jk}\tau'_{\mathfrak{p}_j,k}\} + (b + ib'),$$

where a_{jk}, a'_{jk}, b, b' are arbitrary real constants. The condition that the h periods of this additive function along the paths γ_i vanish is, by (17.2), equivalent to

$$\sum_{j=1}^{r} \sum_{k=1}^{m_j} \left\{ \frac{a_{jk}}{(k-1)!} \Re \frac{d^k w_i}{d\mathfrak{p}^k} + \frac{a'_{jk}}{(k-1)!} \Im \frac{d^k w_i}{d\mathfrak{p}^k} \right\}_{\mathfrak{p}=\mathfrak{p}_j} = 0 \qquad (i = 1, \dots, h).$$

These are h homogeneous linear equations for the $2m$ unknows a_{jk}, a'_{jk}, with real coefficients. Let the rank of this system of equations be $2R$; that is, $2R$ is the number of real linearly independent equations. Then $2A = 2(m + 1 - R)$ is the number of real linearly independent meromorphic functions on $\mathfrak{F}$ which

[17]) Riemann considered only the so-called "general" case ($R = p$ in our notation). The complete result for *integral* divisors $\mathfrak{d}$ is due to G. Roch, Jour. f. Math., **64** (1865) 372–376. Fractional divisors were considered by Klein (*Riemannsche Flächen* I, autograph lectures, Göttingen 1892, pp. 110–111), E. Ritter, Math. Ann., **44** (1894) 314, and Hensel and Landsberg loc. cit., pp. 362–364.

are multiples of $1/\mathfrak{d}$. Let the unknowns in the transposed system of equations, in which the rows and columns are interchanged, be denoted by b_i. With the abbreviation $w = \sum b_i w_i$, the transposed equations demand that

$$\Re\left(\frac{d^k w}{d\,\mathfrak{p}^k}\right)_{\mathfrak{p}_j} = 0, \qquad \Im\left(\frac{d^k w}{d\,\mathfrak{p}^k}\right)_{\mathfrak{p}_j} = 0$$

for $k = 1, \dots, m_j$ and $j = 1, \dots, r$; in other words, that the differential of the first kind dw vanishes with order m_i at the point $\mathfrak{p}_i$ $(1 \le i \le r)$, or that dw is a multiple of $\mathfrak{d}$. The number of differentials which are linearly independent in the real sense and which are multiples of $\mathfrak{d}$ is hence $2B = 2p - 2R$. Thus it appears that $2R$ is even. For the number B of complex linearly independent differentials which are multiples of $\mathfrak{d}$, and the number A of complex linearly independent meromorphic functions which are multiples of $1/\mathfrak{d}$, we have $A = m + 1 - R$, $B = p - R$, and hence

$$A = B + (m + 1 - p).$$

The case in which $\mathfrak{d}$ contains negative exponents may be handled in a similar fashion. Besides the formulae (II_0^n), the reciprocity laws (III) must be used in the process.[18] Thus we have the general *Riemann-Roch theorem.*

Between the number B of linearly independent (in the complex sense) *differentials which are multiples of the arbitrary divisor* $\mathfrak{d}$ *of order m, and the number A of linearly independent functions, uniform on* $\mathfrak{F}$, *which are multiples of the reciprocal divisor* $1/\mathfrak{d}$, *the relation*

$$A = B + (m + 1 - p)$$

holds.

We have seen that if $m > p$, then among the uniform nonconstant functions on $\mathfrak{F}$ there is always a multiple of $1/(\mathfrak{p}_1\mathfrak{p}_2 \dots \mathfrak{p}_m)$. To this we add: if $m \le p$, then *in general*, that is, except for special configurations of the m points $\mathfrak{p}_i$, such a multiple does *not* exist. This again sheds a bright light on the function theoretic significance of the genus p. According to the Riemann-Roch theorem, there exists, among the functions on $\mathfrak{F}$, a nonconstant multiple of

---

[18]) Compare the references to Klein and Hensel-Landsberg in the preceding footnote. E. Ritter, loc. cit., argues as follows. If $f_0 = \mathfrak{d}'/\mathfrak{d}$ is *one* function which is a multiple of $1/\mathfrak{d}$ ($\mathfrak{d}'$ integral), then I obtain *all* such functions f as follows: $f = f_0 f'$, f' a multiple of $1/\mathfrak{d}'$, and thus I come back to the case of an *integral* divisor $\mathfrak{d}'$. But here the proof that [if $B + (m + 1 - p) > 0$] an f_0 exists is lacking.

$1/(\mathfrak{p}_1\mathfrak{p}_2 \ldots \mathfrak{p}_p)$ if and only if there exists a differential dw of the first kind which vanishes at the points $\mathfrak{p}_1, \mathfrak{p}_2, \ldots, \mathfrak{p}_p$ but does not vanish identically. If

$$dw_1^*, dw_2^*, \ldots, dw_p^*$$

is a complex basis for the differentials of the first kind, then it must be possible to determine constants $c_1^*, c_2^*, \ldots, c_p^*$, not all zero, so that

$$c_1^* dw_1^* + c_2^* dw_2^* + \cdots + c_p^* dw_p^*$$

has zeros at those p points. We choose any p distinct points $\mathfrak{p}_1^0, \mathfrak{p}_2^0, \ldots, \mathfrak{p}_p^0$, associated local parameters z_i, and for each $\mathfrak{p}_i^0$ a z_i-disc $K_i: |z_i| \leq a_i$ $(i = 1, 2, \ldots, p)$ such that these discs are disjoint. If, for each choice of the points $\mathfrak{p}_i$ inside K_i, there were a differential of the first kind which vanished at all the points $\mathfrak{p}_i$, then

$$\Delta \equiv \begin{vmatrix} \dfrac{dw_1^*(z_1)}{dz_1}, \ldots, \dfrac{dw_p^*(z_1)}{dz_1} \\ \cdots\cdots\cdots\cdots \\ \cdots\cdots\cdots\cdots \\ \dfrac{dw_1^*(z_p)}{dz_p}, \ldots, \dfrac{dw_p^*(z_p)}{dz_p} \end{vmatrix} = 0$$

would hold identically in $z_1, z_2, \ldots, z_p$ $(|z_i| \leq a_i)$. But this would contradict the linear independence of the differentials dw_i^*. For if we think of z_1 as variable in $\Delta = 0$ and $z_2, \ldots, z_p$ fixed for the moment, then we have a homogeneous linear relation between the differentials $dw_i^*(z_1)$ which is satisfied identically in z_1. Thus the coefficients in this relation, the minors formed from the last $p - 1$ rows of Δ, must all vanish identically in $z_2, \ldots, z_p$. If we repeat this process, applying it to each of the $(p - 1)$-rowed minors and so on, we are reduced to the nonsensical statement that the elements of the last row, that is, the differentials dw_i^* themselves, vanish identically.

To obtain functions on the surface, two ways are open to us: either to combine additive functions linearly with coefficients such that the periods all vanish, or to divide two differentials.[19] By combining both possibilities one can obtain important theorems. We remark first that since the quotient of two differentials is a function, then:

---

[19]) The last method is also available on *open* Riemann surfaces and yields here immediately the proof that there exist functions on the surface which are not constant.

(1) each differential is uniquely determined to within a constant factor by its representative divisor, that is, that divisor whose numerator contains the zeros of the differential with their multiplicities and whose denominator contains the poles; and

(2) the order d of this representative divisor is the same for all differentials; d is the number of zeros of any differential of the first kind, and hence is certainly ≥ 0 for $p > 0$. We shall prove that this important invariant of a Riemann surface *depends only on the genus* p, and, to be sure, according to the simple equation

$$d = 2p - 2.$$

Let $\mathfrak{d}$ and $\mathfrak{e}$ be two divisors, of order m and n, respectively, whose product $\mathfrak{d}\mathfrak{e}$ is the representative of a differential dv on $\mathfrak{F}$: $m + n = d$. Let there be exactly M linearly independent differentials which are multiples of $\mathfrak{d}$, and N which are multiples of $\mathfrak{e}$. Among the uniform functions on $\mathfrak{F}$ we pick out the multiples f of $1/\mathfrak{d}$. If we generate these by combining additive functions, then the Riemann-Roch theorem gives the result that the number of linearly independent f is

$$= m + 1 - p + M.$$

For each f the differential $f\,dv$ is a multiple of $\mathfrak{e}$; that is, we obtain all the functions f in another way, by dividing all differentials which are mutliples of $\mathfrak{e}$ by dv. Considering only the number, we get the relation

$$N = m + 1 - p + M.$$

Analogously,

$$M = n + 1 - p + N.$$

Addition of these two equations yields

$$d = 2p - 2.$$

Subtraction gives the *Brill-Noether reciprocity law:*[20]

$$M - N = \tfrac{1}{2}(n - m).$$

An amusing application of the Riemann-Roch theorem is the so-called

---

[20]) A. Brill and M. Noether, Math. Ann., **7** (1874) 283.

Weierstrass gap theorem.[21] We consider the meromorphic functions on $\mathfrak{F}$ which have a pole only in one point $\mathfrak{q}$; let A_l be the number of linearly independent such functions which become infinite of order at most l at $\mathfrak{q}$. By the Riemann-Roch theorem, $A_l = (l + 1 - p) + B_l$, where B_l is the number of linearly independent differentials of the first kind which have zeros of order at least l at $\mathfrak{q}$. Then $B_0 = p$ and $B_l = 0$ as soon as $l > 2p - 2$; for a differential of the first kind is of order $2p - 2$, and can have no zero of order greater than $2p - 2$ without vanishing identically. Also, either $B_{l+1} = B_l - 1$ or $B_{l+1} = B_l$. The decrease of B_l by one in passing from l to $l + 1$ will occur exactly p times, since B_l decreases from p to 0. Each time this occurs for an exponent l, $A_{l+1} = A_l$; that is, the functions which are infinite of order at most $l + 1$ at $\mathfrak{q}$ coincide with those which are infinite of order at most l. But this says that there exists *no* function which becomes infinite with *precisely the order* $l + 1$ at $\mathfrak{q}$. *Thus there are* p *exponents in the sequence* $e = 1, 2, \ldots$ *for which there exists no uniform function which becomes infinite with precisely the order* e *at* $\mathfrak{q}$. These p gap values of e are all $\leq 2p - 1$.

§ 18. Abel's theorem. The inversion problem

But enough of these considerations, all of which seem to be naturally related to the "partial fraction decomposition" (17.4)! We turn to the "product representation" (17.7) and recall first the definition of a *system of characters.* Such a system is given if to each cover transformation S of $\mathfrak{F}$ there corresponds a number χ_S of absolute value one, and if the correspondence is such that for arbitrary cover transformations S, T, the composition law $\chi_{ST} = \chi_S \chi_T$ holds. It is convenient to use a fixed point $\mathfrak{o}$ on $\mathfrak{F}$ as a center. Then for each point $\mathfrak{a}$ there is a system of integral characters defined by

$$\chi_S(\mathfrak{a}) = \mathrm{ex}\left(\Re \int_{\mathfrak{o}}^{\mathfrak{a}} dw_S\right).$$

In fact, the number $\chi_S(\mathfrak{a})$ is determined uniquely by S and $\mathfrak{a}$; it is not influenced by the choice of the path from $\mathfrak{o}$ to $\mathfrak{a}$. For any divisor (17.8) let the numbers

$$\chi_S(\mathfrak{d}) = \prod_{i=1}^{r} \{\chi_S(\mathfrak{p}_i)\}^{m_i}$$

be introduced as the *integral characters of* $\mathfrak{d}$. The principal result, that we already obtained in the preceding paragraph, may be stated as follows.

---

[21]) K. Weierstrass, *Mathematische Werke,* Vol. **4**, *Vorlesungen über die Theorie der Abelschen Transzendenten,* Berlin, 1902, pp. 224–225.

To each divisor $\mathfrak{d}$, (17.8), *of order* $m = 0$ *there corresponds a multiplicative function* Θ, *unique to within a nonzero constant factor, which is represented by* $\mathfrak{d}$: *at any point* $\mathfrak{p}$, *the order of* Θ *is the exponent associated with* $\mathfrak{p}$ *in the divisor* $\mathfrak{d}$. (This is zero if $\mathfrak{p}$ is not one of the points $\mathfrak{p}_1, \dots, \mathfrak{p}_r$ occurring in the divisor.) *The multipliers* μ_S *of this function* Θ *coincide with the integral characters of* $\mathfrak{d}$.

In particular ($\mu_S = 1$) we obtain the following.

Abel's Theorem (first version). To a divisor $\mathfrak{d}$ *of order* $m = 0$ *there corresponds a uniform meromorphic function on* $\mathfrak{F}$, *represented by* $\mathfrak{d}$, *if and only if the integral characters of* $\mathfrak{d}$ *are all equal to* 1.

Let dv_1 and dv_2 be any two differentials on $\mathfrak{F}$, regular except for poles, and let $\mathfrak{d}_1$ and $\mathfrak{d}_2$ be the divisors representing dv_1 and dv_2. Since $dv_1/dv_2 = \mathfrak{d}_1/\mathfrak{d}_2$ is a function on $\mathfrak{F}$, it follows from Abel's theorem that

$$\chi_S\left(\frac{\mathfrak{d}_1}{\mathfrak{d}_2}\right) = \frac{\chi_S(\mathfrak{d}_1)}{\chi_S(\mathfrak{d}_2)} = 1 \qquad \text{or} \qquad \chi_S(\mathfrak{d}_1) = \chi_S(\mathfrak{d}_2).$$

So we have the following corollary to Abel's theorem.

For all differentials on $\mathfrak{F}$ *or their representative divisors the integral characters have the same values* χ_S^0. *A divisor* $\mathfrak{d}$ *of order* $2p - 2$ *is the representative of a differential if and only if its system of integral characters coincides with the character system* χ_S^0.

Let $\gamma_1, \dots, \gamma_h$ be an integral basis for the closed paths on $\mathfrak{F}$. With each point $\mathfrak{p}$ we associate the h real numbers

$$\xi_i = \Re \int_{\mathfrak{o}}^{\mathfrak{p}} dw_i = \Re\, w_i(\mathfrak{p}) \qquad (i = 1, \dots, h). \tag{18.1}$$

The point $(\xi_1, \dots, \xi_h)$ in the h-dimensional vector space $\Re_h$ is not uniquely determined by $\mathfrak{p}$, but depends to some extent on the path of integration joining $\mathfrak{o}$ to $\mathfrak{p}$. When the path is changed the ξ_i are all altered by integers of the form

$$\{\alpha\}_i = \mathrm{ch}(\alpha, \gamma_i),$$

where α is an arbitrary closed path on $\mathfrak{F}$. The points, with coordinates $\{\alpha\}_1, \dots, \{\alpha\}_h$, corresponding to the various closed α form a *lattice* G which is a sublattice of the unit lattice G_0. The unit lattice contains all points with integral coordinates n_i; G contains all points with coordinates of the form

$$\sum_{j=1}^{h} s_{ij} n_j \qquad (i = 1, \dots, h),$$

where the n_j are arbitrary integers, and the s_{ij} are the coefficients of the characteristic form. Thus $(\xi_1, \dots, \xi_h)$ is determined only modulo G. The number of points of G_0 which are incongruent modulo G is the determinant Δ of the s_{ij}. At the end of the first chapter we claimed, but did not prove, that G coincides with G_0. We shall fill this gap now.

If we write the given divisor $\mathfrak{d}$, of order zero, in the form

$$\frac{\mathfrak{p}_1^0 \cdots \mathfrak{p}_n^0}{\mathfrak{q}_1^0 \cdots \mathfrak{q}_n^0},$$

where the numerator (denominator) contains the zeros (poles) of $\mathfrak{d}$ with the right multiplicities, then Abel's theorem may be stated as follows.

The points $\mathfrak{p}_1^0, \dots, \mathfrak{p}_n^0$ are the zeros, the points $\mathfrak{q}_1^0, \dots, \mathfrak{q}_n^0$ are the poles of a uniform mermorphic function Θ on $\mathfrak{F}$ if and only if the congruences

$$\text{(18.2)} \qquad \mathfrak{R}\, w_i(\mathfrak{p}_1^0) + \cdots + \mathfrak{R}\, w_i(\mathfrak{p}_n^0) \equiv \mathfrak{R}\, w_i(\mathfrak{q}_1^0) + \cdots + \mathfrak{R}\, w_i(\mathfrak{q}_n^0) \qquad (\operatorname{mod} 1)$$

hold for $i = 1, \dots, h$.

The point in $\mathfrak{R}_h$ whose coordinates are the h real numbers on the left-hand sides is uniquely determined modulo G by $\mathfrak{p}_1^0, \dots, \mathfrak{p}_n^0$; likewise the point whose coordinates are the numbers on the right. The necessary and sufficient condition just stated demands that these two points be congruent modulo G_0, and not, as one would naturally expect, that they be congruent modulo G. To conquer this discrepancy between G_0 and G, we look first at the lowest case, $p = 1$.

Here there exists, to within a constant factor, only a single differential dw of the first kind. The values $w(\mathfrak{p})$ corresponding to the different paths of integration from the center $\mathfrak{o}$ to an arbitrary given point $\mathfrak{p}$ form a parallelogrammatic point lattice in the complex w-plane. All lattices thus associated with single points $\mathfrak{p}$ of $\mathfrak{F}$ are related by parallel displacements or have, as we shall say, "the same form and disposition" Λ. Instead of w, one can also use the real parts of w_1 and w_2 $(w_i = w_{\gamma_i})$ corresponding to an integral basis, γ_1, γ_2 of the closed paths. From $dw_1 = \rho_1\, dw$, $dw_2 = \rho_2\, dw$, with the two constants ρ_1 and ρ_2, it follows that $\mathfrak{R}\, dw_1$ and $\mathfrak{R}\, dw_2$ come from $\mathfrak{R}\, dw$ and $\mathfrak{I} dw$ by a linear transformation R with real coefficients. By this transformation the lattice Λ goes into the lattice G spanned by the two vectors

$$(\mathfrak{R}\int_{\gamma_1} dw_1, \mathfrak{R}\int_{\gamma_1} dw_2) = (0, s),$$
$$(\mathfrak{R}\int_{\gamma_2} dw_1, \mathfrak{R}\int_{\gamma_2} dw_2) = (-s, 0).$$

From this we obtain the *discontinuous nature* of Λ. As Λ corresponds to the lattice G, so a finer lattice Λ_0 will correspond to the unit lattice G_0 in the w-plane; the relative density Δ is equal to the square of the integer s.

By regarding each lattice of the form and disposition Λ in the w-plane as a "point," the w-plane is turned into a closed Riemann surface $\mathfrak{T}_\Lambda$. The meromorphic functions on $\mathfrak{T}_\Lambda$ are those analytic functions of w, regular except for poles, which have the same value at all the points of any lattice *(doubly periodic or elliptic functions)*. If two points $\mathfrak{p}$ and $\mathfrak{q}$ of $\mathfrak{F}$ ever corresponded to the same lattice, then by Abel's theorem there would exist a function on $\mathfrak{F}$ which had a single pole of order one at $\mathfrak{p}$ (and a zero at $\mathfrak{q}$); this function would have to map the surface conformally onto the sphere, which is incompatible with $p = 1$. Therefore distinct points $\mathfrak{p}$ always correspond to distinct lattices $w(\mathfrak{p})$. Hence dw can have no zeros; this also follows from our general formula $d = 2p - 2$ for the order of a differential. Thus $\mathfrak{F}$ is mapped topologically (and conformally) onto a part $\mathfrak{G}$ of $\mathfrak{T}_\Lambda$. From Abel's theorem it follows that $\mathfrak{p}$ and $\mathfrak{q}$ coincide, not only when $w(\mathfrak{p}) \equiv w(\mathfrak{q})$ modulo the lattice Λ, but also when this congruence holds modulo Λ_0. Hence the domain $\mathfrak{G}$ cannot contain two points w which are congruent modulo Λ_0 (unless they were congruent modulo Λ). For topological reasons $\mathfrak{G}$ must coincide with all of $\mathfrak{T}_\Lambda$. For as a continuous image of $\mathfrak{F}$, $\mathfrak{G}$ is compact, and its complement $\bar{\mathfrak{G}}$ in $\mathfrak{T}_\Lambda$ is open; on the other hand, $\mathfrak{G}$ itself is open. But the connected set $\mathfrak{T}_\Lambda$ cannot be separated into two open sets $\mathfrak{G} + \bar{\mathfrak{G}}$ unless one of them, here $\bar{\mathfrak{G}}$, is void. Thus it follows that the lattice G coincides with G_0 and $\Delta = s^2 = 1$. The w-plane is the class surface of $\mathfrak{T}_\Lambda$, since the real and imaginary parts of dw are closed differentials on $\mathfrak{T}_\Lambda$; it is also, since it is simply connected, the universal covering surface of $\mathfrak{T}_\Lambda$. For the functions on the base surface $\mathfrak{F}$, w is a *uniformizing variable*; for these functions present themselves as uniform, and even doubly periodic, functions of the variable w in the smooth plane. Two Riemann surfaces of genus one in their normal forms $\mathfrak{T}_{\Lambda'}$ and $\mathfrak{T}_{\Lambda''}$ are conformally equivalent if and only if the lattices of the form Λ' and Λ'' are Euclidean similar.

The elliptic functions solve the "*inversion problem*" in the case $p = 1$. For a surface of arbitrary genus p, the problem is the following. If w_i^* $(i = 1, 2, \ldots, p)$ is a complex basis for the differentials of the first kind, then, for arbitrarily given numbers $\mathrm{F}_1, \ldots, \mathrm{F}_p$, to find points $\mathfrak{p}_1, \ldots, \mathfrak{p}_p$ on $\mathfrak{F}$ so that (for a suitable choice of the paths of integration from $\mathfrak{o}$ to $\mathfrak{p}_l$)

$$(18.3) \qquad \sum_{l=1}^{p} \int_{\mathfrak{o}}^{\mathfrak{p}_l} dw_i^*(\mathfrak{p}) = \mathrm{F}_i \qquad (i = 1, 2, \ldots, p).$$

This inversion problem was formulated by Jacobi, in connection with Abel's theorem. After important preliminary work by Göpel and Rosenhain, this problem was solved in general by Riemann and Weierstrass[22] with the aid of θ-functions, certain transcendental entire functions of the p arguments F_i. The left-hand side of (18.3) depends symmetrically on the p arbitrary point arguments $\mathfrak{p}_1, \ldots, \mathfrak{p}_p$. The p-tubles $(\mathfrak{p}_1, \ldots, \mathfrak{p}_p)$, whose p terms $\mathfrak{p}_l$ vary independently over the surface $\mathfrak{F}$, form a $2p$-dimensional manifold, the p-fold product of $\mathfrak{F}$ with itself. If we identify any two p-tuples which are permutations of each other, then we obtain the *symmetric* product $\{\mathfrak{F} \times \mathfrak{F} \times \cdots \times \mathfrak{F}\}_p$. The right-hand side $(\mathrm{F}_1, \ldots, \mathrm{F}_p)$ in (18.3) is a point of the vector space with p complex coordinates F_i; however, the two points $(\mathrm{F}_1', \ldots, \mathrm{F}_p')$ and $(\mathrm{F}_1, \ldots, \mathrm{F}_p)$ are to be identified whenever the differences $\mathrm{F}_i' - \mathrm{F}_i$ are simultaneous periods of w_i^*, that is,

$$\mathrm{F}_i' - \mathrm{F}_i = \{\alpha\}_i = \int_{\alpha} dw_i^* \qquad (i = 1, \ldots, p)$$

for some closed path α. These simultaneous periods form a discontinuous lattice Λ, and, by identifying any two points which are congruent modulo Λ, the space $\mathfrak{R}$ of vectors $(\mathrm{F}_1, \ldots, \mathrm{F}_p)$ becomes a closed "toruslike" space $\mathfrak{T}_\Lambda$. By our equations (18.3) there is defined an analytic map of the symmetric product $\{\mathfrak{F} \times \ldots \times \mathfrak{F}\}_p$ into $\mathfrak{T}_\Lambda$. One must attempt to prove that the image is all of $\mathfrak{T}_\Lambda$ by means of a topological argument similar to that carried out in

---

[22]) Riemann, *Theorie der Abelschen Funktionen*, Jour. f. Math., **54** (1857) = *Werke*, 2nd ed., pp. 88–142; *Über das Verschwinden der Theta-Funktionen*, Jour. f. Math., **65** (1865) = *Werke*, pp. 212-224. Weierstrass, *Vorlesungen über die Theorie der Abelschen Transzendenten*, Werke Vol. **4**. Stahl, *Theorie der Abelschen Funktionen*, Leipzig 1896. H. F. Baker, *Abel's theorem and the allied theory, including the theory of theta functions*, Cambridge 1897. Prym and Rost, *Theorie der Prymschen Funktionen erster Ordnung*, Leipzig 1911, 2. Teil, 7. Abschnitt. Krazer, *Lehrbuch der Thetafunktionen*, Leipzig 1903.

The principal significance of the inversion problem to us today lies primarily, not in its intrinsic value, but in the splendid developments created by Riemann and Weierstrass in their efforts to solve the problem.

the case $p = 1$. The coincidence of G and G_0 will come out as a corollary. The fact that for $p \geq 2$, this map is not entirely free of singularities (see below) causes certain topological difficulties; for that reason we follow here (in accord with the older literature) a somewhat different route.

We prove the following lemma.

Lemma. If $\mathfrak{f}_i$ $(i = 1, \dots, p)$ are arbitrary given complex numbers, then there exist points $\mathfrak{q}_1^0, \dots, \mathfrak{q}_N^0; \mathfrak{q}_1, \dots, \mathfrak{q}_N$ such that

$$\sum_{\nu=1}^{N} \int_{\mathfrak{q}_\nu^0}^{\mathfrak{q}_\nu} dw_i^* = \mathfrak{f}_i \qquad (i = 1, \dots, p). \tag{18.4}$$

Proof. Let $\mathfrak{q}_1^0, \dots, \mathfrak{q}_p^0$ be p distinct points which are so placed that no differential of the first kind vanishes at all these points (unless it vanishes identically, see p. 138); let z_l be associated local parameters, and let K_l be disjoint z_l-discs about $\mathfrak{q}_l^0$. The equations

$$\sum_{l=1}^{p} \int_{\mathfrak{q}_l^0}^{\mathfrak{q}_l} dw_i^* = \mathfrak{f}_i \qquad (i = 1, 2, \dots, p) \tag{18.5}$$

for the unknowns $\mathfrak{q}_l$ have, if the given values $\mathfrak{f}_i$ are sufficiently small, exactly one solution such that $\mathfrak{q}_l$ (together with the path of integration $\mathfrak{q}_l^0\mathfrak{q}_l$) lies in K_l; this is because of the relation

$$\left| \frac{dw_i^*(z_l)}{dz_l} \right|_{z_1=0, \dots, z_p=0} \neq 0$$

satisfied by the functional determinant.

But if the $\mathfrak{f}_i$ are arbitrary, then choose a positive integer n so large that such a solution $\mathfrak{q}_l$ $(l = 1, \dots, p)$ of the equations (18.5) is possible when one replaces $\mathfrak{f}_i$ on the right by $\mathfrak{f}_i/n$. Then there also exists a solution of the (unchanged) equations (18.4) in which $N = np$ and the sequence $\mathfrak{q}_1, \dots, \mathfrak{q}_N$ $(\mathfrak{q}_1^0, \dots, \mathfrak{q}_N^0)$ contains each of the points $\mathfrak{q}_1, \dots, \mathfrak{q}_p (\mathfrak{q}_1^0, \dots, \mathfrak{q}_p^0)$ n times.

The multiplicative functions Θ, which have previously given characters χ_S as multipliers, form a linear family $\mathfrak{H}_\chi$; that is, any linear combination, with constant coefficients, of functions in this family is another function of the family. From our lemma it follows that *for arbitrarily specified characters χ_S, there exist multiplicative functions*

$$\Theta = \frac{\mathfrak{q}_1 \cdots \mathfrak{q}_N}{\mathfrak{q}_1^0 \cdots \mathfrak{q}_N^0}$$

with the multipliers χ_S. One can now treat the same question for the family $\mathfrak{H}_\chi$ that was answered by the Riemann-Roch theorem for the principal characters $\chi_S = 1$. If $\mathfrak{d}$ is any divisor of order m, then we ask in particular for the functions in $\mathfrak{H}_\chi$ which are multiples of $1/\mathfrak{d}$ and the number A of linearly independent ones. Also here the inequality

$$A \geq m + 1 - p$$

is valid. For one obtains all functions Θ in $\mathfrak{H}_\chi$, which are multiples of $1/\mathfrak{d}$, from *one* function $\Theta_0 = \mathfrak{d}_0$ in $\mathfrak{H}_\chi$, by multiplying Θ_0 by arbitrary *uniform* meromorphic functions which are multiples of $(\mathfrak{d}\mathfrak{d}_0)^{-1}$. Since $\mathfrak{d}\mathfrak{d}_0$, like $\mathfrak{d}$, has order m, the number of linearly independent functions of the last type is $\geq m + 1 - p$.

In particular, this contains the solvability of the inversion problem, for the case $\mathfrak{d} = \mathfrak{o}^p$, assuming that we do not distinguish between congruence modulo G_0 and modulo G. For our theorem says that for arbitrarily given real numbers $\xi_1, \xi_2, \ldots, \xi_p$, there exist points $\mathfrak{p}_1, \ldots, \mathfrak{p}_p$ such that

$$\sum_{l=1}^{p} \Re\, w_i(\mathfrak{p}_l) \equiv \xi_i \qquad (\operatorname{mod} 1)$$

holds for $i = 1, 2, \ldots, 2p$.

Also the formula of the complete Riemann-Roch theorem may be carried over. To do this we must consider, along with the multiplicative functions, the multiplicative differentials which were introduced into the theory by Prym. Such a *multiplicative* or *Prym*[23] *differential* $d\mathsf{T}$ is uniform on $\mathfrak{F}$, analytic, free of essential singularities, and transforms according to

$$d\mathsf{T}(\hat{\mathfrak{p}}\, S) = \chi_S\, d\mathsf{T}(\hat{\mathfrak{p}}),$$

where S is any cover transformation of $\hat{\mathfrak{F}}$, and the χ_S are an initially given system of characters. Every Prym differential may be represented by a divisor $\mathfrak{t}$, whose numerator contains the points, with proper multiplicity, over which $d\mathsf{T}$ has zeros, and whose denominator contains the poles of $d\mathsf{T}$. From the fact

[23]) Jour. f. Math., **70** (1869) 354–362; **71** (1870) 223–236, and 305–315. See the work of Appell cited on page 133, and also the following. Prym and Rost, *Theorie der Prymschen Funktionen erster Ordnung,* Leipzig 1911; R. König, *Zur arithmethischen Theorie der auf einem algebraischen Gebilde existierenden Funktionen,* Ber. d. Verh. Sächs. Ges. Wiss. Leipzig, math.-phys. Kl., **63** (1911) 348–368; O. Haupt, *Zur Theorie der Prymschen Funktionen,* Math. Ann., **77** (1915) 24–64.

that the quotient of a Prym differential and an arbitrary Abelian differential (uniform on $\mathfrak{F}$) is a function Θ with the same multipliers as the Prym differential, it follows that the divisor $\mathfrak{t}$ determines the Prym differential $d\mathsf{T}$ uniquely to within a constant factor. The divisor $\mathfrak{t}$ may be given arbitrarily, provided that its order is $2p - 2$. The multipliers of $d\mathsf{T}$ are $\chi_S(\mathfrak{t}) : \chi_S^0$. Let B be the number of linearly independent Prym differentials, with the multipliers χ_S^{-1}, which are multiples of $\mathfrak{d}$. Then the *generalized Riemann-Roch theorem*,[24]

$$A = B + (m + 1 - p),$$

holds. For B is identical with the number of linearly independent multiples of $\mathfrak{d}\mathfrak{d}_0$ (from which the Prym differentials under consideration arise by division by Θ_0) among the *Abelian* differentials. If we write χ_S^{-1} instead of χ_S, then this relation holds between

> the number B of linearly independent *differentials* which are multiples of $\mathfrak{d}$ and which possess the system χ_S of multipliers, on the one hand, and the number A of linearly independent *functions* which are multiples of the *reciprocal* divisor $1/\mathfrak{d}$ and possess the reciprocal system $1/\chi_S$ of multipliers, on the other hand.

In order to give at least one concrete special case of this theorem, take $\mathfrak{d}$ to be the divisor "1," which contains no point with nonzero exponent. Then, since there are no functions Θ, except constants, without poles, we have $A = 0$ in general; only in the case of principal characters is $A = 1$. Thus, *the Prym differentials of the first kind with given multipliers form a linear family of dimension* $p - 1$ (only in the case of principal characters, the dimension is p).

The inversion problem (18.3) *is always solvable,* as we saw above, *but in general it possesses only one solution.* By the generalized Riemann-Roch theorem, there is more than one solution (infinitely many) only if there exists a Prym differential

$$d\mathsf{T} = \mathfrak{o}^p(\mathfrak{q}_1\mathfrak{q}_2 \cdots \mathfrak{q}_{p-2}) = \mathfrak{t}$$

---

[24]) First formulated by E. Ritter, Math. Ann., **44**, p. 314. (See also the footnote on page. 137). Instead of the symbolism of divisors, Ritter uses the theory of forms on Riemann surfaces, founded by Klein, which makes possible a real representation of divisors with the aid of multiplicative forms. In Math. Ann., **47** (1896) 157–221, aided by the basic theorems introduced into the theory of linear differential equations by Riemann, he generalizes his investigations to *systems* of *n* forms which transform in a homogeneous linear way under the influence of a cover transformation *S*.

with the multipliers χ_S^{-1}, which vanishes with order at least p at the point $\mathfrak{o}$. But then

$$\chi_S = \frac{\chi_S^0}{\chi_S(\mathfrak{t})} = \chi_S^0 : \prod_{i=1}^{p-2} \chi_S(\mathfrak{q}_i).$$

In the $2p$-dimensional manifold of all possible character systems, those obtained from this formula by allowing $\mathfrak{q}_1, \dots, \mathfrak{q}_{p-2}$ to vary independently over the surface $\mathfrak{F}$ form only a $(2p-4)$-dimensional manifold X, which assumes the role of a *singular form* for the inversion problem.[25]

If χ_S is an arbitrary nonprincipal character system, then one can choose the points $\mathfrak{q}_1, \mathfrak{q}_2, \dots, \mathfrak{q}_{p-2}$ such that there exists only one (except for a constant factor) Prym differential

$$d\mathsf{T} = (\mathfrak{q}_1 \cdots \mathfrak{q}_{p-2})(\mathfrak{p}_1 \mathfrak{p}_2 \cdots \mathfrak{p}_p)$$

of the first kind, with the multipliers χ_S, which vanishes at those points. This is according to the theorem that the Prym differentials of the first kind constitute a $(p-1)$-dimensional family. This choice has the consequence that the equation

$$\prod_{l=1}^{p} \chi_S(\mathfrak{p}_l) = \chi_S\left(\chi_S^0 : \prod_{i=1}^{p-2} \chi_S(\mathfrak{q}_i)\right)$$

has *only one solution* $\mathfrak{p}_1, \mathfrak{p}_2, \dots, \mathfrak{p}_p$, or: the character system on the left-hand side of this equation is *not* singular. On the other hand, the fraction in the parentheses on the right belongs to X. Thus the fact that there are only $p-1$ Prym differentials of the first kind is equivalent, if we use the terminology of the inversion problem, to the following theorem.

The singular form X *is never transformed into itself by multiplying all the character systems in* X *by the characters* χ_S *of a fixed system* (except for the trivial case $\chi_S = 1$).

This may suffice to make the relation between the inversion problem and the theory of multiplicative functions and differentials clear. Abel himself proved only that part of the theorem above named for him which states that if $\mathfrak{p}_1^0, \dots, \mathfrak{p}_n^0$ are the zeros and $\mathfrak{q}_1^0, \dots, \mathfrak{q}_n^0$ are the poles of a meromorphic function, then the congruences (18.2) hold. But he proved this part in a

---

[25]) In case $p = 1$, X is not present (p. 143); in case $p = 2$, X consists of a single character system χ_S^0.

sharper form, with congruence modulo G_0 replaced by congruence modulo G.[26] Thus with the inversion already proved, one obtains the following.

Abel's Theorem (second version). The points $\mathfrak{p}_1^0, \ldots, \mathfrak{p}_n^0; \mathfrak{q}_1^0, \ldots, \mathfrak{q}_n^0$ *are the sets of zeros and poles, respectively, of a meromorphic function if and only if, for a suitable choice (independent of w) of the paths from* $\mathfrak{o}$ *to the points* $\mathfrak{p}_\nu^0$ *and* $\mathfrak{q}_\nu^0$, *every Abelian integral w of the first kind satisfies the equation*

$$\sum_{\nu=1}^{n} \int_{\mathfrak{o}}^{\mathfrak{p}_\nu^0} dw = \sum_{\nu=1}^{n} \int_{\mathfrak{o}}^{\mathfrak{q}_\nu^0} dw. \tag{18.6}$$

Let z be any nonconstant meromorphic function on $\mathfrak{F}$. It assumes, as we know, every value the same number of times, say n times. Then the surface $\mathfrak{F}$ may be regarded as an n-sheeted covering surface $\mathfrak{F}_z$ over the z-sphere, if one agrees that a point $\mathfrak{p}$ of $\mathfrak{F}$ lies over the point $z = a$ of the z-sphere if the function z has the value a at $\mathfrak{p}$. This is the conception of a Riemann surface which Riemann himself employed in his works on algebraic functions and their integrals, and one would say today that Abel's argument is most easily understood from this representation. Over those values a, for which the n points $\mathfrak{p}$ at which z assumes the value a are not all distinct, there are certainly fewer than n points of the covering surface. If the function z assumes the finite value a r times at the point $\mathfrak{p}_0$, then $\sqrt[r]{z - a}$ is a local parameter at $\mathfrak{p}_0$; the point $\mathfrak{p}_0$ over a is a branch point of order $r - 1$. If $\mathfrak{U}$ is any neighborhood of $\mathfrak{p}_0$, then there exists a disc $|z - a| \leq \varepsilon$ on the z-sphere such that over every point of this disc (except the center) there are exactly r points of $\mathfrak{F}$ which lie in the neighborhood $\mathfrak{U}$. dz has a zero of order $r - 1$ at $\mathfrak{p}_0$. (Therefore there can be only finitely many such values $z = a$ over which there are

---

[26]) Abel's theorem [developed in splendid simplicity in the short note, *Démonstration d'une propriété générale d'une certaine classe de fonctions transcendentes,* Jour. f. Math., **4** (1829) 200–201 = *Œuvres complètes,* nouvelle édition (1881), **1**, pp. 515–517] is more general, in that it concerns not only integrals of the first kind. A paper of Abel, *Mémoire sur une propriété générale d'une classe très-étendue de fonctions transcendentes,* on this subject, which he submitted to the Paris Academy in 1826, was lost for a long time because of Cauchy's carelessness; it was first published after Abel's death: *Mémoires présentés par divers savants,* **7** (1841) = Abel, *Œuvres complètes,* nouvelle édition (1881), **1**. pp. 145–241. Furthermore, the theorem is contained in the manuscript left by Abel, *Sur la comparaison des fonctions transcendentes, Œuvres complètes,* **2**, pp. 55–66. The converse of Abel's theorem for integrals of the first kind may be read between the lines in Riemann; it was first stated explicitly (without completely adequate proof) by Clebsch, Jour. f. Math., **63** (1864) 198; Clebsch used it for all it was worth in the theory of algebraic curves.

fewer than n points of $\mathfrak{F}_z$.) If z assumes the value ∞ s times at the point $\mathfrak{p}_0$, then $\sqrt[s]{1/z}$ is a local parameter at $\mathfrak{p}_0$ and we have a branch point of order $s-1$; dz has a pole of order $s+1$ at $\mathfrak{p}_0$. We denote the sum of the orders of all the branch points of the surface, its "*branch order*," by V. Then: *the number of zeros – the number of poles of dz* $= \sum(r-1) - \sum(s+1)$, where the first sum on the right is over all points of the surface except those over $z = \infty$, while the second sum is over precisely the points over $z = \infty$. The right-hand side of this equation is

$$= \sum(r-1) + \sum_{\infty}(s-1) - 2\sum_{\infty} s = V - 2n\,,$$

and the left-hand side $= 2p - 2$. Hence

$$V = 2(p + n - 1)\,.$$

The branch order is always even.

Now let dw be an arbitrary differential of the first kind on $\mathfrak{F} = \mathfrak{F}_z$, and let $\mathfrak{p}_1, \ldots, \mathfrak{p}_n$ in general be the points of $\mathfrak{F}$ over the point z on the z-sphere. Then

$$\frac{dW}{dz} = \frac{dw(\mathfrak{p}_1)}{dz} + \cdots + \frac{dw(\mathfrak{p}_n)}{dz}$$

is a differential dW, which is regular everywhere on the z-sphere. But such a differential does not exist, except for $dW = 0$. If we draw any curve on the z-sphere from the south pole $z = 0$ to the north pole $z = \infty$, then the points of $\mathfrak{F}$ over this curve (which sometimes coalesce at branch points) form curves $\gamma_1, \ldots, \gamma_n$, and it follows from integrating dW that

$$\int_{\gamma_1} dw + \cdots + \int_{\gamma_n} dw = 0\,.$$

Each curve γ_ν leads from a zero $\mathfrak{p}_\nu^0$ to a pole $\mathfrak{q}_\nu^0$ of z. If one joins $\mathfrak{o}$ to $\mathfrak{q}_\nu^0$ by a curve $\beta_\nu = \mathfrak{o}\mathfrak{q}_\nu^0$, and denotes the curve $\beta_\nu - \gamma_\nu$ by $\alpha_\nu = \mathfrak{o}\mathfrak{p}_\nu^0$, then we obtain, as claimed, the equation (18.6).

Conversely, the congruences (18.2) follow from these equations, and hence the fact that the multiplicative function

$$\Theta = \frac{\mathfrak{p}_1^0 \cdots \mathfrak{p}_n^0}{\mathfrak{q}_1^0 \cdots \mathfrak{q}_n^0}$$

has the multipliers $\chi_S = 1$ and is uniform on $\mathfrak{F}$.

The coincidence of the lattices G and G_0 now comes out as follows. First, by the lemma there exist points $\mathfrak{q}_1^0, \ldots, \mathfrak{q}_N^0; \mathfrak{q}_1, \ldots, \mathfrak{q}_N$, such that

$$\sum_{\nu=1}^{N} \Re \int_{\mathfrak{o}}^{\mathfrak{q}_\nu} dw_i - \sum_{\nu=1}^{N} \Re \int_{\mathfrak{o}}^{\mathfrak{q}_\nu^0} dw_i = n_i \qquad (i = 1, \ldots, h) \tag{18.7}$$

provided that the n_i are given integers. In this process, definite paths $\beta_\nu^0 = \mathfrak{o}\mathfrak{q}_\nu^0$ and $\beta_\nu = \mathfrak{o}\mathfrak{q}_\nu$ have been chosen. From this it follows that the multiplicative function

$$z = \frac{\mathfrak{q}_1 \cdots \mathfrak{q}_N}{\mathfrak{q}_1^0 \cdots \mathfrak{q}_N^0}$$

is uniform on $\mathfrak{F}$. Hence, by the proof just carried out, paths $\overline{\beta_\nu^0} = \mathfrak{o}\mathfrak{q}_\nu^0$ and $\overline{\beta_\nu} = \mathfrak{o}\mathfrak{q}_\nu$ may be determined so that for these paths the left-hand sides of (18.7) all vanish. Now $\beta_\nu^0 - \overline{\beta_\nu^0} = \alpha_\nu^0$ and $\beta_\nu - \overline{\beta_\nu} = \alpha_\nu$ are closed paths for which

$$\sum_\nu \Re \int_{\alpha_\nu} dw_i - \sum_\nu \Re \int_{\alpha_\nu^0} dw_i = n_i \qquad (i = 1, \ldots, h).$$

The $2N$ closed loops α_ν^0 and α_ν give together a closed path $\alpha = \sum \alpha_\nu - \sum \alpha_\nu^0$ whose intersection numbers $\operatorname{ch}(\alpha, \gamma_i) = \Re \int_\alpha dw_i$, with the curves γ_i of the integral basis, are the given integers n_i.

Analytically one attacks the Jacobi inversion problem as follows. One uses an arbitrary meromorphic function $f(\mathfrak{p})$ on $\mathfrak{F}$ and attempts to determine, not the points $\mathfrak{p}_1, \ldots, \mathfrak{p}_p$ themselves, *but the values of the function f at the points* $\mathfrak{p}_1, \ldots, \mathfrak{p}_p$ from (18.3) in their dependence on $\mathrm{F}_1, \mathrm{F}_2, \ldots, \mathrm{F}_p$. Since the values $f(\mathfrak{p}_1), f(\mathfrak{p}_2), \ldots, f(\mathfrak{p}_p)$ are determined only to within a permutation, it makes better sense to replace them by their *elementary symmetric functions*; that is, by the coefficients of the equation

$$\lambda^p + A_1 \lambda^{p-1} + \cdots + A_p = 0,$$

of degree p whose roots are the numbers

$$\lambda = f(\mathfrak{p}_1), f(\mathfrak{p}_2), \ldots, f(\mathfrak{p}_p).$$

These coefficients A_i, expressed in terms of $\mathrm{F}_1, \mathrm{F}_2, \ldots, \mathrm{F}_p$, are called, following Jacobi's proposal, *Abelian functions*. Except for the singular systems $(\mathrm{F}) = (\mathrm{F}_1, \mathrm{F}_2, \ldots, \mathrm{F}_p)$, which form a subspace of dimension only $(2p - 4)$ in the whole $2p$-dimensional F-space, the Abelian functions are *uniform and regular analytic*. Furthermore, they are $2p$-fold *periodic*. For if γ_l ($l = 1, 2, \ldots, 2p$) is an integral basis for the closed paths on $\mathfrak{F}$ and if

$$a_i^l = \int_{\gamma_l} dw_i^*,$$

then, for every value of the index l, the numbers $a_1^l, a_2^l, \ldots, a_p^l$ are a period system for the Abelian function $A((\mathrm{F}))$:

$$A((\mathrm{F} + a^l)) = A((\mathrm{F})) \qquad \text{identically in} \qquad \mathrm{F}_1, \mathrm{F}_2, \ldots, \mathrm{F}_p.$$

These $2p$ period systems are linearly independent in the following sense: the determinant whose lth row is

$$\mathfrak{R}\, a_1^l, \ldots, \mathfrak{R}\, a_p^l, \mathfrak{I}\, a_1^l, \ldots, \mathfrak{I}\, a_p^l$$

is nonzero. If one carries through explicitly the proof which has been given here of the solvability of the inversion problem, then one obtains the following result: *in every bounded portion*

$$|\mathrm{F}_1| < M, |\mathrm{F}_2| < M, \ldots, |\mathrm{F}_p| < M$$

of the F*-space,* an Abelian function $A((\mathrm{F}))$ may be represented as the quotient of two functions which are regular analytic throughout this bounded portion.[27] The indeterminateness for the singular systems (F) arises because for these values both the numerator and the denominator in the representation vanish. The set of the points of indeterminateness possesses no translations onto itself except for those which obviously come from the periodicity of the Abelian functions.

Furthermore, Riemann and Weierstrass showed, by a much more penetrating analysis, that the Abelian functions, without restriction to a bounded set, may be represented as quotients of transcendental entire functions, the θ-functions. They gave an explicit analytic expression in the form of a rapidly convergent infinite series for the θ-functions. These θ-functions also suffice as a basis for the general theory of the $2p$-fold periodic functions of p independent complex arguments; the theory of Abelian functions is only a special case.[28]

The contents of this and the preceding paragraph show convincingly the ruling role played by the genus p, which is in essence a topological quantity, in the theory of functions and integrals on a closed Riemann surface. We

---

[27]) See Weierstrass, *Werke*, **4**, pp. 451–456.

[28]) See the fourth chapter in Krazer, *Lehrbuch der Thetafunktionen*, Leipzig 1903, where references to the other literature may be found.

were concerned with two trains of thought which we shall state once again with the labels

additive functions, partial fraction decomposition, Riemann-Roch theorem
multiplicative functions, product representation, Abel's theorem.

These two trains of thought permeate the *theory of uniform functions on* $\mathfrak{F}$; from here on this theory is easily completed with the aid of the reciprocity laws of § 16. But the *significance* of this structure shows up in the proper light only when we become familiar with the system of meromorphic functions on a closed Riemann surface from a third point of view, as an *algebraic function field*; this will be done in the next section. And first from this aspect do those functions appear most intimately related to our other interests, the *algebraic* and the *geometric* – in so far as these relate to the theory of algebraic curves in the plane and in spaces of higher dimension.

§ 19. The algebraic function field

The meromorphic functions on a given closed Riemann surface $\mathfrak{F}$ form a *field*; this means precisely that the sum, difference, product, and quotient of two meromorphic functions (only division by 0 is ruled out) is again such a function. As was already done in the last section, one chooses a definite nonconstant meromorphic function z on $\mathfrak{F}$ for the independent variable. Thereby $\mathfrak{F}$ becomes an n-sheeted covering surface $\mathfrak{F}_z$ over the z-sphere, and the uniform meromorphic functions f on $\mathfrak{F}$ become n-valued *algebraic* functions of z. In this fashion there exists an *algebraic function field* belonging to $\mathfrak{F}_z$, which includes the field $k(z) = k$ of rational functions of z. Namely, the meromorphic function f satisfies identically a definite equation

$$(19.1) \qquad F_z(f) = f^n + r_1(z)f^{n-1} + \cdots + r_{n-1}(z)f + r_n(z) = 0,$$

of degree n, in which the $r_i(z)$ are rational functions of z. More precisely, this means the following. Let $\mathfrak{p}^0$ be any point of $\mathfrak{F}$, let t be a local parameter at $\mathfrak{p}^0$, and let $z = z(t)$ and $f = f(t)$ be the expansions of z and f in integral powers of t in the neighborhood of $\mathfrak{p}^0$; then the left-hand side of (19.1) becomes identically zero when one substitutes the power series $z(t)$ and $f(t)$ for z and f, respectively. To derive the equation (19.1), we first exclude the point $z = \infty$ from the z-sphere as well as those points over which there are branch points, and those points over which f has poles. For every other value of z we form the number

$$r_1(z) = f(\mathfrak{p}_1) + f(\mathfrak{p}_2) + \cdots + f(\mathfrak{p}_n),$$

where the sum on the right is over the n points over z. The sum is independent of the ordering of these points. In the neighborhood of a nonexcluded point

$$z_0 : \mathfrak{p}_1^0, \mathfrak{p}_2^0, \dots, \mathfrak{p}_n^0,$$

$f(\mathfrak{p}_1), f(\mathfrak{p}_2), \dots, f(\mathfrak{p}_n)$ become power series in $z - z_0$, and $r_1(z)$ is a regular analytic function at every nonexcluded point. This function cannot have an essential singularity at any excluded point; hence it has only poles. If one uses a local parameter at a point of $\mathfrak{F}_z$ over an excluded value z_0, this may also be verified by elementary calculation. Hence $r_1(z)$ is a *rational* function of z. Similarly, one sees that the other elementary symmetric functions of $f(\mathfrak{p}_1), \dots, f(\mathfrak{p}_n)$ are rational functions of z. The equation (19.1), which is determined uniquely by f, is called the *field equation* of f.

Furthermore, I claim that from the functions g belonging to the function field K associated with $\mathfrak{F}_z$, an f may be chosen such that every g may be expressed rationally in terms of f and z. This f is called a function *determining* the function field. If $\mathfrak{F}_z$ has the n distinct points $\mathfrak{p}_1^0, \dots, \mathfrak{p}_n^0$ over z_0, then it suffices to choose f such that it assumes n distinct values at these points. If $d\tau^k$ $(k = 1, \dots, n)$ is a differential that has a pole only at $\mathfrak{p}_k^0$ and has the principal part $-(z - z_0)^{-2}dz$, then one can set, for example,

$$f = (z - z_0)^2\left(C_1\frac{d\tau^1}{dz} + \cdots + C_n\frac{d\tau^n}{dz}\right),$$

where one chooses any n different constants for $C_1, \dots, C_n$. The field equation of f is then *irreducible*; that is, its left-hand side, the polynomial

$$F_z(u) \equiv u^n + r_1(z)u^{n-1} + \cdots + r_n(z)$$

of degree n in the variable u, cannot be factored into two polynomials $F_z^{(1)}(u)\, F_z^{(2)}(u)$ whose coefficients are also rational functions of z. For assume that this were possible; in a neighborhood of $\mathfrak{p}_1^0$, f may be developed in a power series in the local parameter $z - z_0$ at $\mathfrak{p}_1^0$. Suppose that this power series, replacing u, satisfies the equation $F_z^{(1)}(u) = 0$. Join $\mathfrak{p}_1^0$ to any one of the points $\mathfrak{p}_k^0$ by a curve on $\mathfrak{F}$ whose trace on the z-sphere does not pass through any of the excluded points. Then all the function elements (z, f) along this curve must satisfy the equation $F_z^{(1)}(u) = 0$. Hence $f(\mathfrak{p}_k^0)$ is also a root of the equation $F_{z_0}^{(1)}(u) = 0$; thus it has n distinct roots and must be of degree n. Hence $F_z^{(2)}(u)$ is only of degree 0 in u, and a factorization of the contemplated type does not exist.

To each point $\mathfrak{p}$ of $\mathfrak{F}$ there belongs a function element (z, f) which satisfies the equation $F_z(u) = 0$. For two distinct points these function elements are always different, and the elements belonging to all points $\mathfrak{p}$ exhaust the totality of those which satisfy that equation. In other words: *the totality of those function elements which satisfy the irreducible algebraic equation* $F_z(u) = 0$ *constitutes a single analytic form in the sense of Weierstrass. This analytic form, regarded as a Riemann surface, is conformally equivalent to the given Riemann surface.* The given surface is the Riemann surface which belongs to the algebraic form defined by the equation $F_z(u) = 0$.

In order to express the given element g of the function field K rationally in f, we apply the Lagrange interpolation formula

$$\frac{g(\mathfrak{p}_1)}{u - f(\mathfrak{p}_1)} + \frac{g(\mathfrak{p}_2)}{u - f(\mathfrak{p}_2)} + \cdots + \frac{g(\mathfrak{p}_n)}{u - f(\mathfrak{p}_n)} = \frac{G_z^*(u)}{F_z(u)}.$$

Again $\mathfrak{p}_1, \mathfrak{p}_2, \ldots, \mathfrak{p}_n$ denote the n points of $\mathfrak{F}$ over the point z on the z-sphere. By the same argument as above, one finds that the coefficients of the polynomial $G_z^*(u)$, of degree $n - 1$, are rational functions of z. For all values of z for which $f(\mathfrak{p}_1), \ldots, f(\mathfrak{p}_n)$ are different, it follows that

$$g(\mathfrak{p}_i) = \left[\frac{G_z^*(u)}{F_z'(u)}\right]_{u = f(\mathfrak{p}_i)} \qquad \left\{F_z'(u) = \frac{dF_z(u)}{du}\right\}.$$

Since the polynomial $F_z(u)$ is irreducible over the coefficient field $k(z)$, $F_z(u)$ and $F_z'(u)$ are without common divisor over this field. Therefore the Euclidean division algorithm produces two polynomials, $H_z(u)$ and $L_z(u)$, with coefficients which are rational functions of z, such that

$$H_z(u)\, F_z'(u) + L_z(u)\, F_z(u) = 1\,.$$

If we introduce the polynomial $G_z(u) = H_z(u)\, G_z^*(u)$, then the equation

$$g(\mathfrak{p}) = \{G_z(u)\}_{u = f(\mathfrak{p})}$$

becomes an identity on the surface. Finally, by means of the equation $F_z(u) = 0$, the polynomial $G_z(u)$ may be reduced to one of degree $n - 1$. Hence any element g of our function field has a unique representation of the form

$$g = R_0(z) + R_1(z) f + \cdots + R_{n-1}(z) f^{n-1},$$

where the $R_i(z)$ are rational functions of z. In this sense the quantities $1, f, \ldots, f^{n-1}$ form a *basis* for the algebraic function field K relative to the ground field k of rational functions of z. Hence one calls n the *degree* of K over k.

Here one finds a purely algebraic attack on the idea of a function field. One operates with polynomials $G(u)$ with coefficients in the field k. An irreducible polynomial $F(u)$ of degree n is given. By identifying polynomials $G(u)$ which are congruent modulo $F(u)$, the "*ring*" of polynomials becomes a *field* K, of degree n over the ground field k.[29]

If the independent variable z is replaced by any other function, z^*, on the surface, then there are infinitely many ways of choosing the function f^* on the surface such that all functions may be expressed rationally in terms of z^* and f^*. There is an irreducible algebraic equation $F^*_{z^*}(f^*) = 0$ relating z^* and f^*. Also, z^* and f^* are rational functions of the variables z and f, which are related by $F_z(f) = 0$; conversely, z and f are rational functions of the variables z^* and f^*, which are related by $F^*_{z^*}(f^*) = 0$. Through the *birational transformation* $(z,f) \rightleftarrows (z^*,f^*)$ the equations

$$F_z(u) = 0 \qquad \text{and} \qquad F^*_{z^*}(u^*) = 0$$

turn into each other. The degree of this equation is, naturally, not by any

[29]) The z-plane with finitely many points deleted becomes a simply connected surface if one draws linear cuts from these points to infinity. Riemann's *monodromy problem* is concerned with a far-reaching generalization of algebraic functions (the meromorphic functions on an n-sheeted Riemann surface over the z-sphere). It asks for a system of n functions analytic in the slit plane, and the functions are to be related across the slits by given linear substitutions. (A side condition ruling out essential singularities must be added.) Hilbert, *Grundzüge einer allgemeinen Theorie der Integralgleichungen,* Leipzig 1912, 81–108 and S. J. Plemelj, *Riemannsche Funktionenscharen mit gegebener Monodromiegruppe,* Monatsh. f. Math. u. Physik, Jahrg., **19**, 211–245, demonstrated the solvability of the monodromy problem. Then Robert König developed a complete theory of these "Riemann transcendentals," including the exact analog of the reciprocity laws of §16. See in particular the following works of König: *Die Reduktions- und Reziprozitäts-Theoreme bei den Riemannschen Transzendenten,* Math. Ann., **79** (1918) 76–135 (where one also finds his earlier work developed); *Die Integrale der Riemannschen Transzendenten,* Math. Ann., **80** (1919) 1–28; *Die Elementartheoreme bei den Riemannschen Transzendenten,* Math. Zeit., **15** (1922) 26–65. See also, R. König and M. Kraft, *Über Reihenentwicklung analytischer Funktionen,* Jour. f. Math., **154** (1935) 154–173; ibid., *Über Primfunktionen,* Jour. f. Math., **165** (1931) 96–107. On this foundation is based the work of H. Schmidt, *Über multiplikative Funktionen und die daraus entspringenden Differentialsysteme,* Math. Ann., **105** (1931) 325–380, and, in most recent times, H. Röhrl, *Differentialsysteme, welche aus multiplikativen Klassen mit exponentiellen Singularitäten entspringen,* I, Math. Ann., **123** (1951) 53–75; II, ibid., **124** (1952) 187–218; III, ibid., **125** (1953) 448–466. Further: *Funktionenklassen auf geschlossenen Riemannschen Flächen,* Math. Nachr., **6** (1952) 355–384; *Die Elementartheoreme der Funktionenklassen auf geschlossenen Riemannschen Flächen,* ibid., **7** (1952) 65–84; *Über gewisse Verallgemeinerungen der Abelschen Integrale,* ibid., **9** (1953) 23–44.

means an invariant under birational transformations. But the genus p is invariant under birational transformations.

If there exists a function z on $\mathfrak{F}$ which takes each value just once, then the associated algebraic function field is the field of rational functions of z (and $p = 0$). If there is no function on $\mathfrak{F}$ which takes each value just once, but if there is a function z which takes each value exactly twice, then we may assume that the algebraic equation (which must be quadratic) determining the function field has the form

$$u^2 = (z - e_1)(z - e_2) \dots (z - e_l),$$

where the e_i are all different. These l points and, if l is odd, the point ∞, are branch points of order 1, and hence the genus p is $= (l/2) - 1$ if l is even, and $= (l - 1)/2$ if l is odd. We see that $l = 1$ or 2 leads again to the rational field $p = 0$; $l = 3$ or $l = 4$ gives $p = 1$, which is the elliptic case; if $l > 4$, we get the so-called *hyperelliptic* function fields.

In an arbitrary algebraic function field of genus $p = 1$ there always exists a function with two prescribed poles, therefore a function which assumes each value only twice. In every function field of genus 2 we obtain a function of the same sort by dividing two linearly independent Abelian differentials of the first kind. But, starting with $p = 3$, the hyperelliptic case is no longer the general one.

In the theory of algebraic function fields, two paths are clearly indicated along which one can reach a deeper understanding of the laws governing them. One is that of *abstract algebra*; here the algebraic concepts of field, field extensions, the degree of a field over the ground field, the degree of transcendence, prime place, etc., are paramount; one admits coefficient fields of characteristic other than zero. Here algebraic construction must furnish that which was accomplished in our development by the Dirichlet principle and the method of integration. The other path, that trod by Riemann, is the "*topological*," which we have followed. One may describe Weierstrass' point of view as an algebraic-function-theoretic one lying between these two extremes: explicit construction reigns, but one always operates in the continuum of complex numbers. Similar remarks apply to the "*Kurventheoretiker*" of the German and Italian schools. The concept of an analytic form as a two-dimensional manifold lies close enough here, even if the further step, of regarding the topological properties of this manifold as more primitive than all others, is not completed. Furthermore, it is characteristic of the Riemannian type of development that it is always the Riemann surface, not the analytic form, which is regarded as the given object; the construction of an associated analytic form is a principal component of the problem to be solved. To be sure, in Riemann's treatment itself this point of view does not appear with the complete clarity with which we can now distill it from the works of Prym, Dedekind,[30] C. Neumann, and particularly Klein.[31]

Every closed Riemann surface of genus p may, as we saw, be represented as a multiple-sheeted covering surface over the sphere (with finitely many branch points, and without boundary). There is a great multitude of these "normal forms," even when one normalizes the number of sheets by the condition $n = p + 1$ (which is always possible). A much more fundamental significance attaches to the essentially *unique* normal form of the Riemann surface of arbitrary genus, which is furnished by the theory of uniformization (theory of automorphic functions).

---

[30]) Prym espoused this point of view in his work from 1869 on; here the work of Dedekind, cited on page 33, on elliptic modular functions, from the year 1877, is pertinent.

[31]) See the author's article, *Topologie und abstrakte Algebra als zwei Wege mathematischen Verständnisses,* in Unterrichtsblätter f. Math. u. Naturwiss., **38** (1932) 177–188.

The sketch of the theory of algebraic functions, which we could give here, is very incomplete. From the older literature, which preceded the development of modern abstract algebra, I mention here, besides the works of Riemann, Weierstrass, F. Klein, C. Neumann, H. F. Baker, and Hensel-Landsberg already cited, the following presentations: Clebsch and Gordan, *Theorie der Abelschen Funktionen,* Leipzig 1866; A Brill and M. Noether, *Über die algebraischen Funktionen und ihre Anwendung in der Geometrie,* Math. Ann., **7** (1874) 269–310; Brill and Noether, *Die Entwicklung der Theorie der algebraischen Funktionen,* Bericht der DMV, **3**, Berlin 1894; F. Severi, *Lezione di Geometria algebrica,* Padova 1908 (as representative of the very fruitful Italian school of algebraic geometers). In all of these works the curve-theoretic point of view predominates. A trail-blazing and classical work for the algebraic-arithmetic foundations of the theory is: R. Dedekind and H. Weber, *Theorie der algebraischen Funktionen einer Veränderlichen,* Jour. f. Math., **92** (1882) 181–290, (also contained in H. Weber, *Algebra,* v. III, 2nd edition, Braunschweig 1908, p. 623ff.). Also, F. Klein, *Riemannsche Flächen* I, II, autograph lectures, Göttingen 1892/93. Appell et Goursat, *Theorie des fonctions algébriques,* Paris 1895. H. F. Baker, *Analytic principles of the theory of curves,* Cambridge 1933. The first steps in the domain of algebraic functions with arbitrary fields of constants (including those with prime characteristic) were due to H. Hasse, *Theorie der Differentiale in algebraischen Funktionenkörpern mit vollkommenem Konstantenkörper,* Jour. f. Math., **172** (1943) 55–64, F. K. Schmidt, *Zur arithmetischen Theorie der algebraischen Funktionen,* I, Math. Zeit., **41** (1936) 415, and André Weil, *Zur algebraischen Theorie der algebraischen Funktionen,* Jour. f. Math., **179** (1938) 129–133. A modern overall presentation is that of Claude Chevalley, *Introduction to the theory of algebraic functions of one variable,* Math. Surveys No. 6, Am. Math. Soc., New York 1951.

§ 20. Uniformization

In the theory of uniformization the ideas of Weierstrass and of Riemann grow into a complete unity. With Weierstrass the analytic form (z, u) is described *at each individual point* by a *particular* representation with the aid of a parameter t (the "local parameters"): $z = z(t)$, $u = u(t)$. Certainly Riemann obtains a *global* representation $z = z(\mathfrak{p})$, $u = u(\mathfrak{p})$ of the whole form, but he is forced to regard the parameter $\mathfrak{p}$ as a point on a Riemann surface (not as a complex variable in the usual sense). Uniformization theory is concerned with obtaining a global representation $z = z(t)$, $u = u(t)$ with the aid of a parameter t, the *uniformizing variable,* which varies in a domain of the smooth complex plane. F. Klein and H. Poincaré[32] must be named as the true founders of the *theory of automorphic functions,* to which our problem leads. Their way to the general concepts and results was prepared in the literature by important, but more specialized, investigations of Riemann, Schwarz, Fuchs, Dedekind, Klein, and Schottky. The proof of the possibility of uniformization, based on the concept of a covering surface, was furnished simultaneously in 1907 by P. Koebe and H. Poincaré.[33] From then on, Koebe spent his whole scientific life in studying the problem of uniformization thoroughly from all sides, and with the most varied methods.[34] To him above

---

[32]) For Poincaré, see, besides the numerous notes in the Comptes Rendus for the years 1881/82, the papers in the Acta Mathematica, Vols. **1**, **3**, **4**, **5** (1882/84). For Klein, see the papers in Math. Ann., Vols. **19**, **20**, **21** (1882/83), and also the inclusive presentation: Fricke and Klein, *Vorlesungen über die Theorie der automorphen Funktionen,* Leipzig 1897 to 1912.

[33]) H. Poincaré, *Sur l'uniformisation des fonctions analytiques,* Acta Math., **31** (1907) 1–63. P. Koebe, *Über die Uniformisierung beliebiger analytischer Kurven,* Nachr. d. Ges. Wiss. Göttingen, from 1907 on.

[34]) I give here the titles and places of publication of the larger series of his works; these were frequently announced in shorter notes in the Nachrichten der Gesellschaft der Wissenschaften zu Göttingen and the Sitzungsberichte der Sächsischen Akademie der Wissenschaften zu Leipzig.

(a) *Über die Uniformisierung der algebraischen Kurven,* I, II, III (first proof of the general fundamental theorem of Klein, the iteration procedure), IV (second existence proof for the general canonical uniformizing variable: method of continuity), Math. Ann., **67** (1909) 145–224; **69** (1910) 1–81; **72** (1912) 437–516; **75** (1914) 42–129.

(b) *Über die Uniformisierung beliebiger analytischer Kurven, I. Teil: Das allgemeine Uniformisierungsprinzip; II. Teil: Die zentralen Uniformisierungsprobleme,* Jour. f. Math., **138** (1910) 192–253, and **139** (1911) 251–292.

(c) *Abhandlungen zur Theorie der konformen Abbildung. I. Die Kreisabbildung des allgemeinsten einfach und zweifach zusammenhängenden schlichten Bereichs und die*

all we owe it that today the theory of uniformization, which certainly may claim a central role in complex function theory, stands before us as a mathematical structure of particular harmony and grandeur.

The basic idea of the following proof, to derive the existence of the uniformizing variable from the *Dirichlet principle,* comes from Hilbert.[35]

The uniformizing variable t that we seek should be such that it is suitable as a local parameter at each point of the given surface $\mathfrak{F}$. Therefore it must be a uniform function, regular analytic except for poles of order one, on the universal covering surface $\tilde{\mathfrak{F}}$. If we seek that function t which possesses the strongest uniformizing power, then we will attempt to determine t such that it assumes distinct values at any two distinct points of the surface $\tilde{\mathfrak{F}}$; then it will map $\tilde{\mathfrak{F}}$ one-to-one and conformally onto a domain of the t-sphere. Then not only the functions on the base surface $\mathfrak{F}$ can be represented as uniform functions of t; but also the much larger class of functions (which are in general infinitely many-valued on $\mathfrak{F}$) which arise from any function element on $\mathfrak{F}$ which can be continued, without branching, along all paths in $\mathfrak{F}$. And since $\tilde{\mathfrak{F}}$ (in contrast to $\mathfrak{F}$) is simply connected, the possibility of such a

---

Ränderzuordnung bei konformer Abbildung, Jour. f. Math., **145** (1915) 177–223. II. *Die Fundamentalabbildung beliebiger mehrfach zusammenhängender schlichter Bereiche nebst einer Anwendung auf die Bestimmung algebraischer Funktionen zu gegebener Riemannscher Fläche,* Acta Math., **40** (1916) 251–290. III. *Der allgemeine Fundamentalsatz der konformen Abbildung nebst einer Anwendung auf die konforme Abbildung der Oberfläche einer körperlichen Ecke,* Jour. f. Math., **147** (1917) 67–104. IV. *Abbildung mehrfach zusammenhängender schlichter Bereiche auf Schlitzbereiche,* Acta Math., **41** (1918) 305–344. V. *Fortsetzung,* Math. Zeit., **2** (1918) 198–236. VI. *Abbildung mehrfach zusammenhängender Bereiche auf Kreisbereiche. Uniformisierung hyperelliptischer Kurven,* Math. Zeit., **7** (1920) 235–301.

(d) *Riemannsche Mannigfaltigkeiten und nichteuklidische Raumformen, acht Mitteilungen;* Sitzber. Akad. Berlin, 1927, 164–196; 1928, 345–384 and 385–442; 1929, 414–457; 1930, 304–364 and 505–541; 1931, 506–534; 1932, 249–284. Also the prize winning monograph of 1920, first published in Acta Math., **50** (1927) 27–157, *Allgemeine Theorie der Riemannschen Mannigfaltigkeiten.*

See also Koebe's survey lectures: *Über ein allgemeines Uniformisierungsprinzip,* Int. Math. Cong. Rome, Atti (1909) 25–30; *Referat über automorphe Funktionen und Uniformisierung,* Jahresber. DMV **21** (1912) 157–163; *Methoden der konformen Abbildung und Uniformisierung,* Math. Congr. Bologna, Atti, **3** (1928) 195–203.

Of the more recent literature, the following should be noted especially: B. L. van der Waerden, *Topologie und Uniformisierung der Riemannschen Flächen,* Ber. Sächs. Ak. Wiss. Leipzig, math.-phys. Kl., **93** (1941) 147–160; and R. Nevanlinna's book, *Uniformisierung,* Berlin 1953.

[35]) *Zur Theorie der konformen Abbildung,* Nachr. Ges. Wiss. Göttingen, 1909, 314–323.

map does not contradict the analysis-situs properties of $\tilde{\mathfrak{F}}$. We may well drop now the basic assumption of the preceding sections, that $\mathfrak{F}$ be closed; this assumption would not simplify anything in uniformization theory.

We obtain the desired uniformizing variable simply by applying the Dirichlet principle, not to $\mathfrak{F}$, but to the universal covering surface $\tilde{\mathfrak{F}}$. We choose a point $\mathfrak{o}$ on $\tilde{\mathfrak{F}}$ with the local parameter ζ; then we construct on $\tilde{\mathfrak{F}}$, with the aid of the Dirichlet principle, the potential function U which is regular everywhere on $\tilde{\mathfrak{F}}$ except at $\mathfrak{o}$, which behaves like $\mathfrak{R}(1/\zeta)$ at $\mathfrak{o}$, and with the following properties.

(1) The Dirichlet integral of U over all of $\tilde{\mathfrak{F}}$, except for an arbitrarily small ζ-disc about $\mathfrak{o}$, is finite,

(2) For every continuously differentiable function w on $\tilde{\mathfrak{F}}$, with finite Dirichlet integral, which vanishes in a neighborhood of $\mathfrak{o}$, the variation $D(U, w)$ satisfies

$$D(U,w)=0.$$

U generates a differential $d\tau$ on $\tilde{\mathfrak{F}}$; since $\tilde{\mathfrak{F}}$ is simply connected, this must be the differential of a certain function

$$\tau = U + iV,$$

whose real part coincides with U, which is regular analytic everywhere except at $\mathfrak{o}$, and which has a pole of order one at $\mathfrak{o}$. Then τ is a uniformizing variable of the type we seek. The proof of this fact follows in a very elegant fashion with the aid of the following deduction, due to Koebe.[36]

We prove first the following fact.

If V_0 is any real constant, then the points of $\tilde{\mathfrak{F}}$ at which $V > V_0$ form a single domain; similarly for the points where $V < V_0$.

Now $1/\tau$ is a local parameter at $\mathfrak{o}$; let $K_0 : |1/\tau| \leq a_0$, be a $(1/\tau)$-disc about $\mathfrak{o}$. Let $\mathfrak{E}(V_0)$ be the closed set on $\tilde{\mathfrak{F}}$ consisting of the points where $V = V_0$; then certainly only two of the domains determined by $\mathfrak{E}(V_0)$ have points in K_0. If our claim were false, then among the domains determined by $\mathfrak{E}(V_0)$ there would be one, say $\mathfrak{G}$, that did not penetrate the neighborhood K_0 of $\mathfrak{o}$. Now let $\phi(u)$ and $\psi(u)$ be any two real functions, defined and continuously

---

[36]) *Über die Hilbertsche Uniformisierungsmethode*, Nachr. Ges. Wiss. Göttingen, 1910, 61–65.

differentiable for all real values u. We form the following function w on the surface $\mathfrak{F}$:

$$w = \begin{cases} \phi(U)\psi(V), & \text{for all points of } \mathfrak{G}, \\ 0, & \text{for all points not in } \mathfrak{G}. \end{cases}$$

This function is everywhere continuously differentiable, provided that

$$\psi(V_0) = 0, \qquad \psi'(V_0) = 0 \qquad \left(\psi' = \frac{d\psi}{du}\right).$$

In the neighborhood K_0 of $\mathfrak{o}$, w vanishes identically. If $\mathfrak{p}$ is any point in $\mathfrak{G}$ and if $z = x + iy$ is a local parameter at $\mathfrak{p}$, then

$$\begin{aligned} \frac{\partial w}{\partial x} &= \phi'(U)\psi(V)\frac{\partial U}{\partial x} + \phi(U)\psi'(V)\frac{\partial V}{\partial x} \\ &= \phi'(U)\psi(V)\frac{\partial U}{\partial x} - \phi(U)\psi'(V)\frac{\partial U}{\partial y}, \\ \frac{\partial w}{\partial y} &= \phi'(U)\psi(V)\frac{\partial U}{\partial y} + \phi(U)\psi'(V)\frac{\partial U}{\partial x}. \end{aligned}$$

If ϕ, ϕ', ψ, and ψ' are bounded functions, then the Dirichlet integral of w over all of $\mathfrak{F}$ will be finite. Then, under the given conditions, $D(U, w) = 0$. Now

$$\frac{\partial w}{\partial x}\frac{\partial U}{\partial x} + \frac{\partial w}{\partial y}\frac{\partial U}{\partial y} = \phi'(U)\psi(V)\left[\left(\frac{\partial U}{\partial x}\right)^2 + \left(\frac{\partial U}{\partial y}\right)^2\right].$$

If we choose ϕ and ψ such that ϕ' and ψ are positive for all values of their argument (except for $u = V_0$ in the case of ψ), then we have a contradiction.[37]

From the fact proved, and the simple connectivity of $\mathfrak{F}$, one can draw conclusions on the behavior of τ.

(1) *$d\tau$ has no zeros.* If at a point $\mathfrak{p}_0$ on $\mathfrak{F}$, where $\tau = \tau_0$, $d\tau = 0$, then not $\tau - \tau_0$, but $(\tau - \tau_0)^{1/r}$ (r an integer ≥ 2) would be a local parameter at $\mathfrak{p}_0$; let us assume $r = 2$; the proof for larger r is analogous. I set $\tau - \tau_0 = \sigma^2$

---

[37]) All the required conditions are satisfied if we use the functions

$$\alpha(u) = \operatorname{arctg} u \quad \left(-\frac{\pi}{2} < \alpha(u) < \frac{\pi}{2}\right), \qquad \beta(u) = \frac{u^2}{1 + u^2}:$$

$$\phi(u) = \alpha(u), \qquad \psi(u) = \beta(u - V_0).$$

and draw in the complex σ-plane a disc K, with center $\sigma = 0$, so small that it is, via the function σ, the conformal image of a certain neighborhood of the point $\mathfrak{p}_0$ on $\mathfrak{F}$. I take four points $\mathfrak{p}_1, \mathfrak{p}_2, \mathfrak{q}_1, \mathfrak{q}_2$ in K, as indicated in Fig. 9, which are the vertices of a cross formed by two linear segments, α and β, crossing at the origin. At $\mathfrak{p}_1$ and $\mathfrak{p}_2$, $V > V_0$; at $\mathfrak{q}_1$ and $\mathfrak{q}_2$, $V < V_0$. I think of this figure transferred to the surface $\mathfrak{F}$; then I can join $\mathfrak{p}_1$ and $\mathfrak{p}_2$ by a curve α', at every point of which $V > V_0$; likewise, I can join $\mathfrak{q}_1$ and $\mathfrak{q}_2$ by a curve β' on which $V < V_0$. The intersection number of the two closed curves, $\alpha + \alpha'$ and $\beta + \beta'$, is thus $= 1$. But this contradicts the fact that $\alpha + \alpha'$, as a curve on the simply connected surface $\mathfrak{F}$, must be homologous to zero.

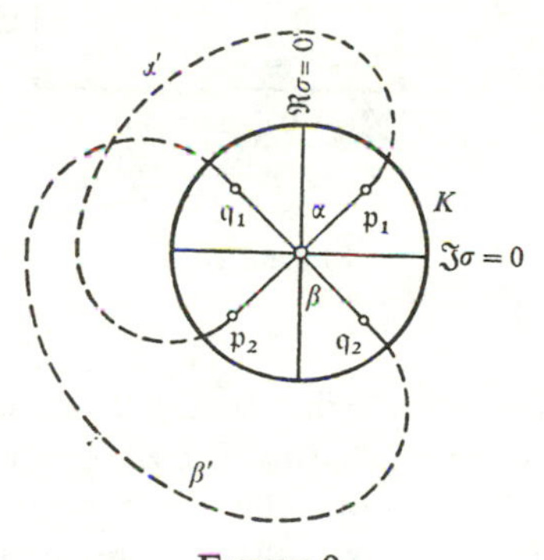

FIGURE 9

(2) If I say that a point of the surface $\mathfrak{F}$, at which τ has the value τ_0, lies over the point τ_0 of the τ-sphere, then $\mathfrak{F}$ becomes a covering surface $\mathfrak{F}_\tau$ of the τ-sphere or the τ-plane. By what we have proved under (1), this covering is unbranched. There is just one point, $\mathfrak{o}$, over $\tau = \infty$. I follow the line $V = V_0$, starting at $\tau = \infty$, in the τ-plane in the direction from smaller to larger values of U; on $\mathfrak{F}$, the point $\tilde{\mathfrak{p}}$, starting at $\mathfrak{o}$, traces a curve over this line in the τ-plane. If I do not run into any boundary before returning to ∞, then $\tilde{\mathfrak{p}}$ describes a closed curve $\tilde{\gamma}$, for there is only the single point $\mathfrak{o}$ over ∞; then $\tilde{\gamma}$ covers the line $V = V_0$ in the τ-plane simply. But if I meet an obstruction, then I obtain a curve $\tilde{\gamma}_1$ on $\mathfrak{F}_\tau$, which covers a certain piece $U < U_1$ of the line $V = V_0$ *simply*. Then I follow the line $V = V_0$, now from larger to smaller values of U, and obtain a curve $\tilde{\gamma}_2$ on $\mathfrak{F}_\tau$ which covers a piece $U > U_2$ of the line simply. The sum $\tilde{\gamma}_1 + \tilde{\gamma}_2$ forms a curve $\tilde{\gamma}$ through $\mathfrak{o}$, nonclosed, and without end on $\mathfrak{F}_\tau$. In either case, the curve $\tilde{\gamma}$ separates the surface $\mathfrak{F}_\tau$, since it is simply connected, into two domains, $\mathfrak{G}'$ and $\mathfrak{G}''$. If,

outside of $\tilde{\gamma}$, there were points of $\tilde{\mathfrak{F}}$ at which $V = V_0$, say in $\mathfrak{G}'$, then there would be points in $\mathfrak{G}'$ at which $V < V_0$ and points at which $V > V_0$. The point set $\mathfrak{E}(V_0)$ would then determine at least three domains. Therefore $\tilde{\gamma}$ exhausts the points at which $V = V_0$. Thus, *as a covering of the* τ*-sphere,* $\tilde{\mathfrak{F}}$ *is everywhere at most two sheeted. A value* $U_0 + iV_0$ *will certainly occur once and only once on* $\tilde{\mathfrak{F}}$*, if* $V = V_0$ *is a closed curve on* $\tilde{\mathfrak{F}}$*.*

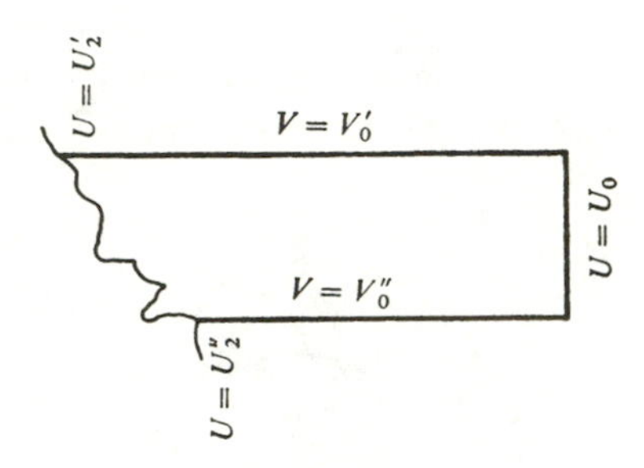

FIGURE 10

We still want to prove carefully that $\tilde{\gamma}$ must always separate the surface $\tilde{\mathfrak{F}}$. For this purpose we construct a two-sheeted, unbranched, and unlimited covering surface over $\tilde{\mathfrak{F}}$ as follows. Cut $\tilde{\mathfrak{F}}$ along $\tilde{\gamma}$, take two copies of this cut surface and identify the edges of the slits criss-cross. Put abstractly, this amounts to the following. To each point $\tilde{\mathfrak{p}}$ of $\tilde{\mathfrak{F}}$ we associate two points "over it," $\tilde{\mathfrak{p}}^1$ and $\tilde{\mathfrak{p}}^2$. If $\tilde{\mathfrak{p}}_0$ is a point not on $\tilde{\gamma}$, $\tau_0 = \tau(\tilde{\mathfrak{p}}_0)$, and K an arbitrary $(\tau - \tau_0)$-disc which does not intersect $\tilde{\gamma}$, then the points $\tilde{\mathfrak{p}}^1$ (with upper index 1) which lie over the interior points $\tilde{\mathfrak{p}}$ of K constitute a "neighborhood" of $\tilde{\mathfrak{p}}_0^1$; those points $\tilde{\mathfrak{p}}^2$ over the same points $\tilde{\mathfrak{p}}$ constitute a "neighborhood" of $\tilde{\mathfrak{p}}_0^2$. On the other hand, if $\tilde{\mathfrak{p}}_0$ is on $\tilde{\gamma}$, let K denote an arbitrary $(\tau - \tau_0)$-disc. Those points $\tilde{\mathfrak{p}}^1$ whose trace points $\tilde{\mathfrak{p}}$ lie inside K and satisfy the condition $V \geq V_0$, together with all points $\tilde{\mathfrak{p}}^2$ lying over inner points $\tilde{\mathfrak{p}}$ of K which satisfy the condition $V < V_0$, shall constitute a "neighborhood" of $\tilde{\mathfrak{p}}_0^1$. The neighborhood of $\tilde{\mathfrak{p}}_0^2$ is defined analogously. In the last case all the points in K at which $V = V_0$ certainly belong to $\tilde{\gamma}$; hence this definition of the concept of neighborhood is in agreement with all the demands to be made of such a definition. If $\tilde{\gamma}$ does not separate $\tilde{\mathfrak{F}}$, then it is clear that the manifold just defined also satisfies the condition that any two of its points can be joined by a continuous curve. But the existence of such a covering surface would contradict the fact that $\tilde{\mathfrak{F}}$ is simply connected.

The last step in the proof is the demonstration of the following theorem.

There exists at most one real number V_0 such that the associated line $V = V_0$ on $\mathfrak{F}$ is nonclosed.

For if there were two such lines $\tilde{\gamma}'$ and $\tilde{\gamma}''$, say $V = V_0'$ and $V = V_0''$, then one makes use of a *closed* line $U = U_0$ completely contained in K_0 (U_0 is chosen large enough). Let $U > U_2'$ be one of the pieces $\tilde{\gamma}_2'$ of the curve $\tilde{\gamma}'$; let $U > U_2''$ be the piece $\tilde{\gamma}_2''$ of $\tilde{\gamma}''$. Over the heavily drawn path (Fig. 10, the three segments) in the τ-plane there is a curve in $\mathfrak{F}_\tau$, which covers the path simply, and has no ends on $\mathfrak{F}_\tau$. This curve separates the simply connected $\mathfrak{F}$ into two domains; let $\mathfrak{G}$ be that one of the two domains which does not contain the point $\mathfrak{o}$. Again we set

$$w = \begin{cases} \phi(U)\psi(V) & \text{inside } \mathfrak{G} \\ 0 & \text{outside } \mathfrak{G}. \end{cases}$$

For this function to be continuously differentiable on $\mathfrak{F}$, we must have

$$\phi(U_0) = \phi'(U_0) = 0, \qquad \begin{aligned} \psi(V_0') &= \psi'(V_0') = 0, \\ \psi(V_0'') &= \psi'(V_0'') = 0. \end{aligned}$$

Let ϕ, ϕ', ψ, ψ' be bounded and let ϕ' (except for $u = U_0$) and ψ (except for $u = V_0'$ or V_0'') be positive.[38] Then we get a contradiction of the equation $D(U, w) = 0$. Thus we have proved the following.

> *τ maps the surface $\mathfrak{F}$ one-to-one and conformally either onto the complete sphere (case 1)*

or

> *onto the sphere with one point τ_0 removed (case 2)*

or

> *onto the sphere with a slit $V = V_0$, $U_1 \leq U \leq U_2$ removed (case 3).*

We replace τ by a somewhat different uniformizing variable t. To be sure, $t = \tau$ in case 1. In the second case we arrange it so that the point omitted from the sphere is the point at ∞: set $t = 1/(\tau - \tau_0)$; t maps the covering surface onto the whole plane (without the point at ∞). In the third case we first apply an entire linear transformation so that the slit is given by

$$V = 0, \qquad (-1 \leq U \leq +1).$$

---

[38]) All the demands are satisfied by [see footnote 37]:

$$\phi = \alpha(u - U_0)\beta(u - U_0), \qquad \psi = \beta(u - V_0)\beta(u - V_0'').$$

Then, with the aid of the formula

$$\tau = \frac{1}{2}\left(t + \frac{1}{t}\right),$$

the slit τ-sphere is mapped conformally onto the interior of the unit disc $|t| < 1$ in the t-plane.

The construction we have described of the uniformizing variable t occurs in two clearly separate steps. First, the surface $\mathfrak{F}$ solves the problem in so far as it belongs to analysis situs. Then the function theoretic theorem, *every simply connected Riemann surface may be mapped conformally onto a domain on the sphere,* applied to $\mathfrak{F}$ gives the uniformizing variable. By a slight modification of the argument, one can also show that every *planar* surface may be mapped conformally onto a domain on the sphere. For nonplanar surfaces this is impossible, for reasons of analysis situs: there will not exist even a *topological* map onto a domain on the sphere. Every uniformizing variable belonging to $\mathfrak{F}$ (unbranched relative to $\mathfrak{F}$) will map a certain unbranched, unlimited, planar covering surface $\bar{\mathfrak{F}}$ over $\mathfrak{F}$ conformally onto a plane domain. One assigns two uniformizing variables which map the same covering surface $\bar{\mathfrak{F}}$ onto subdomains of the plane to the same *class* $\{\bar{\mathfrak{F}}\}$. The determination of all uniformizing variables requires then the solution of two problems.

(1) The analysis-situs problem: to determine all unbranched unlimited planar covering surfaces $\bar{\mathfrak{F}}$ of a given surface $\mathfrak{F}$.

(2) The conformal mapping problem: to find all possible conformal maps of a planar surface $\bar{\mathfrak{F}}$ onto a plane domain.

That the last is always possible in at least one (and hence in infinitely many) way (in other words, that every class of uniformizers $\{\bar{\mathfrak{F}}\}$ conceivable under the analysis-situs condition of planarity actually exists in the function theoretic sense) is the content of *Koebe's general uniformization principle.*[39] As a matter of fact, this principle goes even further; it includes not only the uniformizing variables which are unbranched relative to $\mathfrak{F}$, but also the great multitude of branched uniformizing variables. But without question, the uniformizing variable which we have set up and denoted by t carries more fundamental significance than any other uniformizing variable.

---

[39]) See particularly P. Koebe, *Über die Uniformisierung beliebiger analytischer Kurven, erster Teil; Das Allgemeine Uniformisierungsprinzip,* Jour. f. Math., **138** (1910) 192–253.

§ 21. Riemann surfaces and non-Euclidean groups of motions. Fundamental regions. Poincaré Θ-series

To what extent is the uniformizing variable t determined by the properties stated at the end of the last section? That is, in how many ways can one map the surface $\mathfrak{F}$ conformally onto the sphere, the plane, or the unit disc? This question obviously comes down to the following: in how many ways can one map the sphere, plane, or disc conformally onto one of these three domains? To begin with, it is clear that the sphere, since it is closed, can be mapped onto neither the plane nor the disc. Also, a conformal map of the plane onto the disc is impossible. If the function $t^*(t)$ transformed the t-plane conformally onto the unit disc in the t^*-plane, then it would be an entire function whose absolute value would be bounded by one. By Liouville's theorem, there is no such function (except for the constants, which make no sense here). Furthermore, the following simple theorems hold.

(CASE 1). *The set of conformal maps of the sphere* (represented in the usual fashion by a complex variable) *onto itself is the set of linear transformations.*

(CASE 2). *The conformal maps of the complex plane onto itself are given precisely by the entire linear transformations.*

(CASE 3). *Likewise, the open unit disc can be mapped conformally onto itself only by linear transformations.*

Case 1. We have to show that if the t-sphere is mapped conformally onto the t^*-sphere, $t^* = t^*(t)$, so that $t = 0$ goes into $t^* = 0$ and $t = \infty$ goes into $t^* = \infty$, then the map must be of the form $t^* = ct$, where c is a constant. Now $1/t^*$ is regular at the north pole ($t = \infty$) of the t-sphere and has a zero there; so t/t^* is regular there. Since the relation $t \to t^*$ is one-to-one, t^* vanishes only at $t = 0$, where it has a zero of order one; hence t/t^* is regular on the complete t-sphere, and is thus a constant.

Case 2. Again we may assume that $t = 0$ is mapped into $t^* = 0$. Next one has to prove that

$$\lim \frac{1}{t^*} = 0 \qquad \text{for} \qquad \lim \frac{1}{t} = 0. \tag{21.1}$$

The circle $|t^*| = R^*$ corresponds to a closed curve $\mathfrak{C}$ in the t-plane; let R_0 be the maximum distance of a point on $\mathfrak{C}$ from the origin, and let R be an arbitrary number $> R_0$. In the disc $|t| \leq R$, $|t^*|$ assumes its maximum, which must be $> R^*$, at some boundary point, say $t = t_0$. The image $\mathfrak{K}^*$, of the circle $|t| = R$, in the t^*-plane cannot meet the circle $|t^*| = R^*$; for the circle $|t| = R$ and the curve $\mathfrak{C}$, of which those two curves are the images, do not meet in the t-plane. Since the point $t^*(t_0)$ on $\mathfrak{K}^*$ has an absolute value $> R^*$, we must have $|t^*| > R^*$ for all points on $\mathfrak{K}^*$. That is, $|t| > R_0$ implies $|t^*| > R^*$, and this is the claim (21.1). Hence $1/t^*(t)$ is regular at the north pole of the t-sphere and has a zero there. The rest of the argument is exactly the same as in case 1.

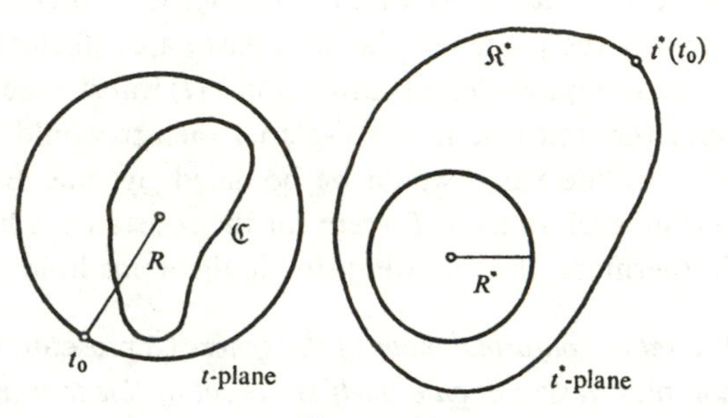

FIGURE 11

We settle case 3 with the aid of the so-called Schwarz lemma.[40] Again we may assume that $t = 0$ goes into $t^* = 0$. If I consider the regular function t^*/t in the disc $|t| \leq q\ (< 1)$, it must attain its maximum absolute value on the boundary, and hence this maximum is $< 1/q\ (|t^*| < 1, |t| = q)$. Since I can choose q arbitrarily close to 1, $|t^*/t| \leq 1$ must hold at all points of the open unit disc. In a similar fashion, $|t/t^*| \leq 1$, and hence $|t^*/t| = 1$. This is possible only if t^*/t is a constant of modulus one.

With this the stated claims are demonstrated: *the uniformizing parameter t is always uniquely determined to within a linear transformation.* The linear transformations which map the open unit disc onto itself have the form

$$t^* = \frac{(a+ib)t + (c-id)}{(c+id)t + (a-ib)} \tag{21.2}$$

$$\{a, b, c, d \text{ real}\,;\ \text{determinant}\,(a^2+b^2) - (c^2+d^2) = 1\}.$$

In particular, since the cover transformations of $\mathfrak{F}$ are one-to-one conformal

---

[40]) H. A. Schwarz, *Gesammelte Abhandlungen*, Vol. II, pp. 109–111. The application of this lemma to our problem goes back to Poincaré, Acta Math., **4** (1884) 231–232.

maps of $\bar{\mathfrak{F}}$ onto itself, these cover transformations must correspond to linear transformations in the t-plane. Thus there corresponds to the group of cover transformations a certain isomorphic group Γ of linear transformations. No transformation of the group Γ (except the identity) can have a *fixed point* in the image domain (sphere, plane, or disc); a fixed point is a point that goes into itself under the transformation. Since every linear transformation of the sphere has a fixed point, there are no cover transformations in case 1, except the identity. Then $\bar{\mathfrak{F}}$ is the same as $\mathfrak{F}$, and $\mathfrak{F}$ is identical, as a Riemann surface, with the sphere. *Case 1 can arise only when the given Riemann surface $\mathfrak{F}$ is equivalent to the sphere.*

The points of $\bar{\mathfrak{F}}$ over one point of $\mathfrak{F}$ appear in the map as a system of points t equivalent under Γ. Such a system Σ has the property that any point of the system can be carried into any other by some transformation of Γ; and also, the image of any point of Σ by any transformation of Γ is another point of Σ (see p. 28). The uniform functions on $\mathfrak{F}$, regular except for poles, appear, when expressed as functions of t, as *automorphic functions* attached to Γ. That is, functions $z(t)$ which are invariant under the transformations of Γ:

$$z\left(\frac{\alpha t + \beta}{\gamma t + \delta}\right) = z(t),$$

where $t^* = (\alpha t + \beta)/(\gamma t + \delta)$ is any transformation in the group Γ. The group Γ must be *discontinuous*; that is, a system of equivalent points under Γ can never have a limit point in the image domain. If we define a Riemann surface $\mathfrak{F}_\Gamma$ by taking the "points" of $\mathfrak{F}_\Gamma$ to be the systems of equivalent points under Γ and by carrying the angular measure in the t-plane directly over to $\mathfrak{F}_\Gamma$, then $\mathfrak{F}_\Gamma$ is equivalent, as a Riemann surface, to the given $\mathfrak{F}$.

$\mathfrak{F}_\Gamma$ is the most appropriate normal form into which every Riemann surface can be brought.

In case 2, Γ must consist of entire linear transformations which have no finite fixed point. The only transformations satisfying this condition are the translations $t' = t + \alpha$. A discontinuous group of translations must be of one of the following types.[41]

(1) Γ contains only the identity. Then $\bar{\mathfrak{F}}$ is identical with $\mathfrak{F}$, and $\mathfrak{F}$ is equivalent to the *plane* (the "*simply punched*" *sphere,* that is, the sphere without the north pole).

---

[41]) That the three cases listed exhaust the possibilities follows from the process of the adaptation of a (two-dimensional) number lattice relative to a sublattice. See page 89 of this book or G. Kowalewski, *Die komplexen Veränderlichen und ihre Funktionen,* Leipzig 1911, pp. 57 f.

(2) Γ consists of the iterations of a single translation

$$\Gamma : t^* = t + n\alpha \qquad (n = 0, \pm 1, \pm 2, \ldots);$$

then clearly $\mathfrak{F}$ is equivalent to the infinitely long right *circular cylinder,* and hence, by the Mercator projection, to the *doubly punched sphere* (the sphere without north and south poles).

(3) Γ will consist of the transformations

$$t^* = t + m\alpha + n\beta \qquad \begin{pmatrix} m = 0, \pm 1, \pm 2, \ldots \\ n = 0, \pm 1, \pm 2, \ldots \end{pmatrix},$$

where α and β are translations in different directions. Then $\mathfrak{F}$ is closed; dt is a uniform differential, regular everywhere on $\mathfrak{F}$, without any zeros, and hence $\mathfrak{F}$ is a *closed Riemann surface of genus one.* Every such surface – whose type (but not its most general form) is the torus – admits in fact a uniformization by the integral of the first kind, by means of which the universal covering surface is mapped onto the plane.

Aside from the few exceptions listed above ($\mathfrak{F}$ = sphere, simply or doubly punched sphere, closed surfaces of genus one), case 3, in which the image domain is the *open unit disc,* always occurs. Since in general the periphery of the unit disc is a natural boundary (cut) for the automorphic functions of t which represent the functions on the base surface $\mathfrak{F}$, one calls t a *cut-circle uniformizing variable* [*Grenzkreis-Uniformisierende*].

The group Γ is not determined uniquely by the given surface. For t may be replaced by any variable t' obtained from t by a linear transformation T_0 which preserves the open unit disc. This transforms Γ into the group

$$\Gamma' = T_0^{-1} \Gamma T_0 .$$

We introduce the following terminology. Any point t of the open unit disc is a "point of $\mathfrak{E}$." For the "straight lines in $\mathfrak{E}$" we take the circular arcs in $\mathfrak{E}$ which are orthogonal to the unit circle (the diameters of the unit disc are included in these "straight lines"). Any linear transformation preserving the open unit disc is a "motion in $\mathfrak{E}$"; two point sets in $\mathfrak{E}$ are "congruent" if one can be carried onto the other by a "motion." The usual angular measure is retained. Then the complete *Bolyai-Lobatschefsky geometry* holds for these "points" and "straight lines." That is the geometry whose axioms are the same as the axioms of Euclidean geometry, except that the parallel axiom is omitted.[42] So we may call $\mathfrak{E}$ the *non-Euclidean plane*. The complex variable t (restricted by the inequality $|t| < 1$) is to be regarded as a coordinate representing the points in the Lobatschefskian plane, in the same way that the rectangular Cartesian coordinates x, y, or their complex fusion $z = x + iy$, represent the points of the Euclidean plane. Any linear transformation of t which preserves the unit disc provides another equally valid coordinate system for the Lobatschefskian plane. In this interpretation, Γ now appears as a group of motions of the non-Euclidean plane; or, more precisely, as a *representation of such a group* by means of a definite coordinate t. If we replace t by t', obtained from t by a linear transformation preserving the unit disc, then the new group Γ' thus obtained is to be regarded as a *different* representation of the *same* group of motions of the non-Euclidean plane (with the aid of another coordinate t'). There are no rotations in Γ, that is, no motions of $\mathfrak{E}$ which leave some point of $\mathfrak{E}$ fixed.

Thus to every Riemann surface (aside from the four exceptions already listed) there corresponds a single uniquely determined discontinuous group Γ *of motions of the Lobatschefskian plane; and* Γ *contains no rotations. Two Riemann surfaces are conformally equivalent (as Riemann expressed it, belong to the same class or are realizations of one and the same ideal Riemann surface) if and only if the associated groups of non-Euclidean motions are congruent in the sense of Lobatschefskian geometry. Conversely, to every rotation-free discontinuous group of non-Euclidean motions there corresponds a definite class of Riemann surfaces.* (The surface $\mathfrak{F}_\Gamma$ constructed on page 169 will serve as a representative of this class.[43])

To present a discontinuous group Γ of motions of the n-E plane[44] in a more visual fashion, one uses, following Klein and Poincaré, a simple gapless covering of the plane by n-E congruent regions *(fundamental regions)* which are permuted by the motions of Γ. A dissection of this sort, which is as simple as possible, is provided by the following considerations.

Definition of n-E distance. Let t_1 and t_2 be any two distinct points in the

---

[42]) The model of the plane non-Euclidean geometry which presents itself here is intimately related to the model, discovered by Klein in the year 1871, which is based on Cayley's metric. Klein, *Über die sogenannte Nicht-Euklidische Geometrie,* Math. Ann., **4** (1871) 573–625. See also Fricke and Klein, *Vorlesungen über die Theorie der automorphen Funktionen,* Vol. I, Leipzig 1897, pp. 3–59. The book of Bonola gives a good orientation on non-Euclidean geometry: *Non-Euclidean geometry: a critical and historical study of its developments,* English translation by H. S. Carslaw, Dover Pubs., 1955 (originally published 1911).

[43]) See the final remarks in Koebe's first communication *Über die Uniformisierung beliebiger analytischer Kurven,* Göttinger Nachrichten, 1907, 209–210.

[44]) n-E = non-Euclidean.

open unit disc; join them by a n-E line, that is, a circular arc in the Gaussian t-plane which intersects the unit circle orthogonally at t_{∞_1} and t_{∞_2}. This circular arc is uniquely determined by t_1 and t_2. The points on it are to come in the sequence t_{∞_1}, t_1, t_2, t_{∞_2} (Fig. 12). The cross-ratio

$$d(t_1 t_2) = \frac{t_1 - t_{\infty 2}}{t_1 - t_{\infty_1}} : \frac{t_2 - t_{\infty 2}}{t_2 - t_{\infty_1}}$$

has a positive real value > 1, and its real logarithm

$$r(t_1, t_2) = \lg d(t_1, t_2)$$

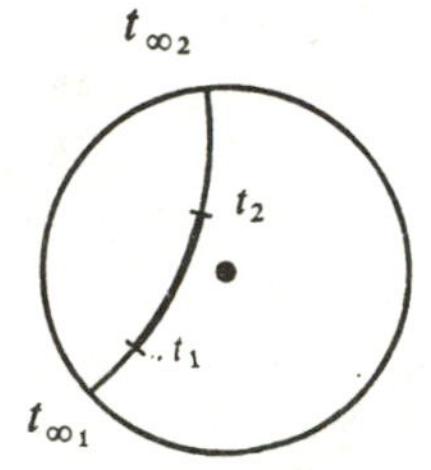

Fig. 12. Non-Euclidean segment.

is positive. $r(t_1, t_2)$ is invariant under all linear transformations preserving the unit disc. The n-E segment $t_1 t_2$ is n-E congruent to another such segment $t_1^* t_2^*$ if and only if $r(t_1, t_2) = r(t_1^*, t_2^*)$. Furthermore, if the three points t_1, t_2, t_3 lie in that order on an n-E line, then

$$r(t_1, t_2) + r(t_2, t_3) = r(t_1, t_3).$$

Because of these two facts, we shall speak of the number $r(t_1, t_2)$ as the *non-Euclidean distance* between the points t_1 and t_2. Now we are in a position to measure not only angles but also distances in the n-E plane.

If $\boldsymbol{\Sigma}_0$ is a system of equivalent points under Γ, if t_0 is a single point of $\boldsymbol{\Sigma}_0$, and t an arbitrary point of the n-E plane, then because of the discontinuity of Γ there are only a finite number of points of $\boldsymbol{\Sigma}_0$ whose distance[45] from t does not exceed an arbitrary given number R. Thus among the points of $\boldsymbol{\Sigma}_0$ there is one or several (but certainly only a finite number) of points t_h such

[45]) The word "distance," and all other geometric expressions, are to be understood, until further notice, in the n-E sense.

that the distance $r(t, t_h)$ is the *smallest* distance of t from any point of Σ_0. Then, as I shall express it briefly, t lies closest to t_h. If the distance $r(t, t_h)$ is *strictly less* than the distance to all other points of Σ_0, then I can find a neighborhood of t such that every point of this neighborhood lies closest to t_h. Let t_h be any point of Σ_0; I collect all those t which lie closest to t_h into a point set $\mathfrak{P}_h$ with "center" t_h. The resulting sets $\mathfrak{P}_h$, with all centers in Σ_0, have disjoint interiors and cover the whole n-E plane. The transformation of Γ which carries t_0 into t_h carries $\mathfrak{P}_0$ into $\mathfrak{P}_h$; hence $\mathfrak{P}_h$ is n-E congruent to $\mathfrak{P}_0$. To every point t there corresponds a point of $\mathfrak{P}_0$ which is equivalent to t under Γ; two distinct *interior* points of $\mathfrak{P}_0$ are never equivalent. Thus $\mathfrak{P}_0$ has the properties of a fundamental region. Also, $\mathfrak{P}_0$ is convex; that is, if t' and t'' are any two points of $\mathfrak{P}_0$, then all points of the n-E segment $t't''$ joining t' and t'' belong to $\mathfrak{P}_0$. The perpendicular bisector $\mathfrak{g}_h$ of the segment t_0t_h is the locus of points equidistant from t_0 and t_h; to every t_h distinct from t_0 we obtain such a line $\mathfrak{g}_h$. The boundary of the closed set $\mathfrak{P}_0$ is composed of segments of these lines; the interior of $\mathfrak{P}_0$ consists of those points which, for every $\mathfrak{g}_h$, lie on the same side of $\mathfrak{g}_h$ as t_0. All points of the line $\mathfrak{g}_h$ have a distance $\geq \frac{1}{2}r(t_0, t_h)$ from t_0, and there are only a finite number of the centers t_h whose distance from t_0 does not exceed an arbitrarily given $2R$. Hence it follows that only a finite number of segments on the boundary of $\mathfrak{P}_0$ has a distance $\leq R$ from the principal center t_0. Thus $\mathfrak{P}_0$ is a convex polygon, with a finite or infinite number of sides, whose vertices (that is, those points which are equidistant from three or more points of the system Σ_0) have no limit point in the finite.

Following Fricke, $\mathfrak{P}_0$ is called a *normal polygon* belonging to the group Γ. The sides of the normal polygon are associated in pairs; either side of a pair can be carried onto the other by some motion in Γ. Two sides with a common vertex are never associated; for such a common vertex would have to be a fixed point of the motion carrying one side onto the other. These motions carrying one side of $\mathfrak{P}_0$ onto another (that is, the motions carrying $\mathfrak{P}_0$ onto a fundamental region $\mathfrak{P}_h$ which abuts $\mathfrak{P}_0$ along some side of $\mathfrak{P}_0$) generate the group Γ. That is, every motion of Γ may be obtained by iterating and composing those particular motions. One proves this as follows. Join an arbitrary point t_h of Σ_0 to t_0 by a polygonal path which avoids the vertices of the polygonal decomposition $\{\mathfrak{P}_h\}$. Now observe the succession of polygons in this decomposition through which the polygonal path passes. Suppose there are in all e polygons $\mathfrak{P}_h$ equivalent to $\mathfrak{P}_0$ (including $\mathfrak{P}_0$ itself) which meet at the vertex *1* of $\mathfrak{P}_0$. Then there are exactly e centers to which *1* lies closest, and there are e points of $\mathfrak{P}_0$ which are equivalent to *1* under Γ; or, in the terminology of Poincaré, which constitute a *cycle* of vertices of the polygon $\mathfrak{P}_0$. The sum of the angles in $\mathfrak{P}_0$ at the vertices of such a cycle is the full angle 1 (= 360°).

The Riemann surface $\mathfrak{F}_\Gamma$ (p. 169) belonging to the group Γ is obviously closed if and only if there exists a positive number R such that for every point in the n-E plane there is at least one equivalent point under Γ whose n-E distance from the center t_0 is $\leq R$. Then every point has a distance $\leq R$ from the closest point in the system Σ_0; in particular, $\mathfrak{P}_0$ is completely contained in the n-E disc of radius R about t_0. Since the vertices of $\mathfrak{P}_0$ cannot have a limit point in the finite, there are, in this case, only a finite number of such vertices: $\mathfrak{P}_0$ *has finitely many sides.* Let the number of sides of $\mathfrak{P}_0$, which must be even because of the pairing of the sides, be $2s$; let c denote the number of distinct vertex cycles of $\mathfrak{P}_0$, so that c is also the sum of the angles in $\mathfrak{P}_0$. We wish to show that the genus p of this closed Riemann surface may be computed from s and c by the simple formula

$$2p = s - c + 1. \tag{21.3}$$

Since every cycle contains at least three vertices, we have

$$3c \leq 2s, \qquad c \leq \frac{2s}{3}, \qquad s - c \geq \frac{s}{3},$$

and hence

$$\frac{s}{3} \leq 2p - 1.$$

Therefore $12p - 6$ is an upper bound for $2s$, the number of sides or vertices of a normal polygon.

We choose $t_0 = 0$. The (n-E as well as Euclidean) linear segments joining 0 to the vertices of $\mathfrak{P}_0$ separate $\mathfrak{P}_0$ into $2s$ triangles. This provides a triangulation of $\mathfrak{F}_\Gamma$ into $2s$ triangles with $3s$ edges and $c + 1$ vertices. Then, from the most basic formula of combinatorial topology, the genus p is in fact determined by the equation

$$2p - 2 = 3s - 2s - (c + 1) = s - c - 1.$$

In this book we have adopted the methodological attitude of avoiding boundaries as much as possible, and we have operated with coverings by overlapping neighborhoods rather than with dissections into simple pieces of surface. The appeal here to combinatorial topology would introduce a foreign element. Therefore, instead of this topological proof of the equation (21.3), I prefer a function theoretic proof. It runs as follows.

We take a meromorphic differential dz on the closed surface $\mathfrak{F}_\Gamma$. As we know, its order is $2p - 2$. We shall assume for the moment that none of the zeros or poles of dz lie on the sides of $\mathfrak{P}_0$. Consider the analytic function dz/dt of t in the region $\mathfrak{P}_0$. Being a convex polygon, $\mathfrak{P}_0$ is simply connected, and one of the simplest consequences of the Cauchy integral theorem in function theory says that the increase in the azimuth of dz/dt, in tracing the boundary of $\mathfrak{P}_0$, is equal to the number of zeros less the number of poles of dz/dt in the interior of $\mathfrak{P}_0$; hence it is $= 2p - 2$. To determine the increase of the azimuth of dz/dt in tracing the boundary, consider a side $\sigma = t_1 t_2 = AB$ of $\mathfrak{P}_0$ and its equivalent side $\sigma^* = t_1^* t_2^*$, where the second comes from

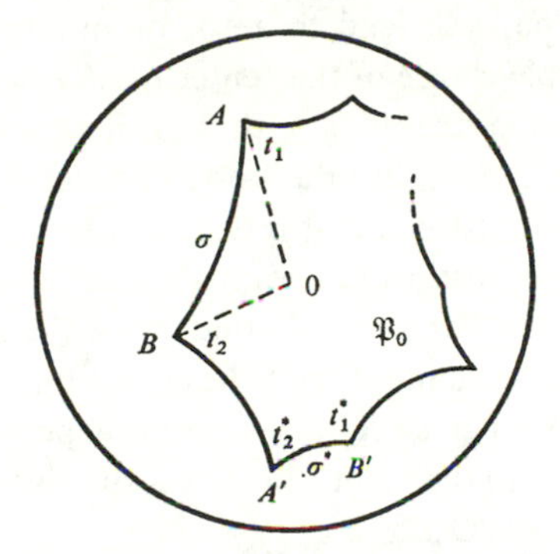

Fig. 13. Determination of the genus from the normal polygon.

the first by the substitution $t^* = tS$ of the group Γ. This substitution S carries the polygon $\mathfrak{P}_0$ into an equivalent one which abuts on $\mathfrak{P}_0$ along the side σ^*. If, in tracing $\mathfrak{P}_0$ in the positive sense, the segment σ is traced from t_1 to t_2, then the segment σ^* will be traced in the sense $t_2^* t_1^* = A'B'$. Since dz is invariant under the substitution S, the sum of the contributions of these two segments to the increase of the azimuth is equal to the difference

$$\left[\mathrm{az}\frac{dt^*}{dt}\right]_{t_1}^{t_2} = -\,[\mathrm{az}(dt^*)]_{t_2^*}^{t_1^*} - [\mathrm{az}(dt)]_{t_1}^{t_2}$$
$$= \sphericalangle(AB) + \sphericalangle(A'B').$$

Here $\sphericalangle\,\sigma = \sphericalangle\,(AB)$ denotes the negative of the angle (measured in Euclidean fashion) through which the tangent to $\sigma = AB$ turns when this side is traced in the positive sense from A to B. This angle is positive and $= \frac{1}{2}$ minus the sum of the angles of the n-E triangle OAB. In this way we get, for the desired increase in the azimuth of dz/dt when the boundary of $\mathfrak{P}_0$ is traced in the positive sense, the value

$$s - \{1 + \text{sum of the angles of the polygon } \mathfrak{P}_0\} = s - (1 + c),$$

and the equation claimed, $2p - 2 = s - c + 1$, follows.

It is inherent in the situation that one cannot completely avoid bordered regions in a dissection into equivalent fundamental regions, but the boundaries here are simple enough to be grasped easily.

If dz has zeros or poles on the boundary of $\mathfrak{P}_0$, then one can proceed in one of two ways. Either one cuts out disc neighborhoods of these zeros and poles and later lets the radii tend to zero, or one avoids this unpleasant eventuality by a suitable choice of the center t_0. If $t = a$ is a zero or pole of dz which lies on the mid-line of $t_0 t_0^*$, that is, if t_0 and $t_0^* = t_0 S$ have the same distance from a, then t_0 has the same distance from a as from aS^{-1}. So one need only choose t_0 such that it does not lie on any of the mid-lines associated with a pair of distinct but equivalent zeros or poles of dz; then it is certain that no point a lies on the boundary of the normal polygon with center t_0. Since only a finite number of these mid-lines come within an n-E distance of the origin $\leq$ a given R, this is certainly possible. Finally, one can bring the point $t = t_0$ into the point $t = 0$ by a suitable n-E motion and thus make $t = 0$ the center of $\mathfrak{P}_0$.

The dissection of the n-E plane into normal polygons furnishes not only a visualization of the n-E group of motions, but also a means of constructing such groups directly.[46]

In the group Γ of motions of the non-Euclidean plane (or in the associated manifold $\mathfrak{F}_\Gamma$) we meet the purest embodiment of the concept of a Riemann surface, freed of all accidental properties. To crown the whole development of this part of Riemannian function theory, the solution of the following problem would be desired. *Given a Riemann surface in its normal form* (that is, the associated group of motions of the non-Euclidean plane), *to express*

---

[46]) Besides the normal polygons, also the "canonical polygons," whose theory was developed by Fricke (Fricke–Klein, *Vorlesungen über die Theorie der automorphen Funktionen,* Leipzig, 1897–1912) are of great importance. With their aid, Fricke succeeded in formulating and proving rigorously the theorem which had already been stated by Riemann: the Riemann surfaces of genus p (> 1) form a $(6p - 6)$-dimensional manifold. These matters are intimately connected with the "method of continuity," by means of which Klein and Poincaré attempted to prove, simultaneously at the beginning of the 1880's, the uniformizibility of algebraic forms. See in particular P. Koebe, Math. Ann., **75** (1914) 42–129.

each of the uniform or many-valued unbranched functions on the surface by means of a closed analytic formula in terms of the coordinate t of the points in the Lobatschefskian plane. The Θ-series introduced by Poincaré[47] constitute an important attack on the solution of this problem. If the given group Γ which preserves the open unit disc consists of the substitutions

$$t_i^* = tS_i = \frac{\alpha_i t + \beta_i}{\gamma_i t + \delta_i} \qquad (\alpha_i\delta_i - \beta_i\gamma_i = 1)$$
$$(i = 0, 1, 2, 3, \dots; S_0 \text{ the identity}),$$

then, for example,

$$\Theta(t) = \sum_{i=0}^{\infty} \frac{1}{(\alpha_i t + \beta_i)(\gamma_i t + \delta_i)^3}, \qquad \Theta'(t) = \sum_{i=0}^{\infty} \frac{1}{(\alpha_i t + \beta_i)^3(\gamma_i t + \delta_i)}$$

are Poincaré Θ-series. To establish the uniform absolute convergence of these series, take an ordinary disc κ of radius a: $|t - t_0| \leq a$ about an arbitrary point t_0 ($|t_0| < 1$) so small that the equivalent discs κS_i are mutually disjoint. The surface integral, J_i, of

$$\left|\frac{dt_i^*}{dt}\right|^2 = \frac{1}{|\gamma_i t + \delta_i|^4}$$

over κ is the (Euclidean) area of κS_i; consequently, $\Sigma_{i=0}^{\infty} J_i$ is convergent. If the Euclidean distance from t to t_0 satisfies $|t - t_0| \leq a/2$, then by (12.16)

$$\left(\frac{3a}{4}\right)^2 \pi \left|\frac{dt_i^*}{dt}\right|^2 \leq J_i.$$

Therefore

$$\sum \frac{1}{|\gamma_i t + \delta_i|^4}$$

is uniformly convergent in the neighborhood of $t = t_0$. Also,

$$\lim_{i=\infty} \left|\frac{\alpha_i t + \beta_i}{\gamma_i t + \delta_i}\right| = 1$$

holds uniformly for $|t| \leq q < 1$, for the equivalent points t_i^* condense only on the boundary of the unit disc; the uniform absolute convergence of the two series, Θ, and Θ' follows. At $t = 0$, and at all equivalent points under Γ,

---

[47]) Poincaré, *Mémoire sur les fonctions fuchsiennes*, Acta Math., **1** (1882) 207.

Θ has a pole of order 1; otherwise Θ is regular. At the same points Θ' has poles of order 3 and is regular everywhere else. From the equation

$$\Theta(t)(dt)^2 = \sum_{i=0}^{\infty} \frac{(dt_i^*)^2}{t_i^*}$$

it follows that $\Theta(t)\,(dt)^2$ goes into itself under an arbitrary transformation S_i of the group

$$\Theta(tS_i) = (\gamma_i t + \delta_i)^4\,\Theta(t).$$

Similarly,

$$\Theta'(tS_i) = (\gamma_i t + \delta_i)^4\,\Theta'(t).$$

Hence $\Theta(t)/\Theta'(t)$ is an automorphic function relative to the group Γ; it is not identically zero, but is has a zero of order 2 at $t = 0$ (and at every equivalent point). In short, it is a uniform nonconstant function, without essential singularities, on the ground surface $\mathfrak{F}_\Gamma$.

The examples of Θ-series given here are of dimension -4. Poincaré succeeded (by using n-E instead of Euclidean area) in proving[48] the absolute convergence of Θ-series of dimension $-r$, for $r > 2$. Once a Riemann surface is given in its normal form, one might hope, on the basis of Poincaré's attack, that one could succeed in deriving all the theorems on the existence of functions and integrals on the surface with the aid of analytic formulae. This would be the analog of what has long been possible in the case $p = 1$ with the theory of elliptic functions. Perhaps one would prefer then to construct Riemannian function theory as follows: with the aid of the Dirichlet principle or a competing method derive, not the existence of functions and integrals, but only the existence of the cut-circle uniformizing variable; then construct the functions and differentials by explicit formulae with this uniformizing variable as argument. The recent extensive investigations of H. Petersson are aimed in this direction. The difficulty lies in the limit passage from $r > 2$ to $r = 2$. The structural form of the Poincaré series does not appear to furnish an adequate foundation for this process; Petersson reaches his goal only through a new device of regarding the Θ-series as eigenfunctions of certain

[48]) For a brief formulation of this proof see H. Petersson: Abh. Math. Sem. Univ. Hamburg, **16** (1948) 127–130; and, in more detail, *Über den Bereich absoluter Konvergenz der Poincaréschen Reihen,* Acta Math., **80** (1948) 23–63. See also: *Über die Transformationsfaktoren der relativen Invarianten linearer Substitutionsgruppen,* Monatsh. f. Math., **53** (1949) 17–41.

functionals.[49] In spite of the tremendous advance attained by Petersson, it seemed advisable to me to retain, in this book, the independent construction of all functions and differentials by means of the Dirichlet principle.

§ 22. The conformal mapping of a Riemann surface onto itself

We say that a group of motions of the n-E plane contains *infinitesimal transformations* if there exist motions in the group which differ arbitrarily little from the identity. A discontinuous group of motions, say Γ, certainly contains no infinitesimal transformations. Also the converse of this theorem is true.

An n-E group of motions is discontinuous if and only if it contains no infinitesimal motions.

To prove this we use, as before, the complex coordinate t. In general, a linear transformation $T: t \to t^*$ has two fixed points on the t-sphere, τ' and τ''; then the transformation may be written[50]

$$\frac{t^* - \tau''}{t^* - \tau'} = \mu \frac{t - \tau''}{t - \tau'}; \tag{22.1}$$

the constant μ is called the *multiplier* of T. If T preserves the open unit disc, then there are two possibilities: (1) *either* τ' and τ'' both lie on the circumference of the unit disc, and μ is positive (*hyperbolic* transformation); *or* (2) τ' and τ'' are inverse points relative to the unit circle, and τ' is interior to the same; then $|\mu| = 1$ (*elliptic* transformation). In the n-E plane, T is a rotation about τ', and the real number ϕ, determined to within an additive integer by the equation $\mu = \mathrm{ex}(\phi)$, is the angle of rotation of T.

The *parabolic* transformations insert themselves as the trasnitional case between the hyperbolic and elliptic transformations. A parabolic transfor-

---

[49]) *Automorphe Formen als metrische Invarianten,* Vols. I and II, Math. Nachr., **1** (1948) 158–212 and 218–257; *Über Weierstrass-Punkte und die expliziten Darstellungen der automorphen Formen von reeller Dimension,* Math. Zeit., **52** (1949) 32–59.

[50]) If $\tau'' = \infty$, this becomes

$$\frac{1}{t^* - \tau'} = \mu \frac{1}{t - \tau'}.$$

mation has only *one* fixed point, τ'. If it preserves the open unit disc, then it has the form

$$\frac{1}{t^* - \tau'} = \frac{1}{t - \tau'} + \frac{ib}{\tau'} \qquad (|\tau'| = 1, b \text{ real}). \tag{22.2}$$

We base our proof on the following lemma.

Lemma. Let

$$t_n, t_n^* \qquad (n = 1, 2, 3, \ldots)$$

be two sequences of points, in the n-E plane, which converge to the same point t_0 ($|t_0| < 1$). Let T_n be an n-E motion which carries t_n into t_n^*. If there exists a neighborhood of t_0 which contains no fixed point of any of the motions T_n ($n = 1, 2, 3, \ldots$), then T_n converges to the identity.

Let τ_n' ($|\tau_n'| \leq 1$) and τ_n'' be the two fixed points of T_n (they may coincide); then there is a positive number l such that

$$|t_0 - \tau_n'|, \qquad |t_0 - \tau_n''| > \frac{3l}{2}$$

holds for all n. If I set

$$|t_n - t_n^*| = \varepsilon_n,$$

then I can assume that

$$|t_n - t_0|, \qquad |t_n^* - t_0| < \frac{l}{2}, \qquad \text{thus} \qquad \varepsilon_n < l$$

for all n. Then every one of the four differences

$$|t_n - \tau_n'|, \qquad |t_n - \tau_n''|, \qquad |t_n^* - \tau_n'|, \qquad |t_n^* - \tau_n''|$$

is $> l$. If T_n is not parabolic, then it has the form

$$\frac{t^* - \tau_n''}{t^* - \tau_n'} = \mu_n \frac{t - \tau_n''}{t - \tau_n'}. \tag{22.3}$$

If I set $t = t_n$, and thus $t^* = t_n^*$, in here, I get

$$\mu_n = \frac{t_n^* - \tau_n''}{t_n - \tau_n''} : \frac{t_n^* - \tau_n'}{t_n - \tau_n'} = \left(1 + \frac{t_n^* - t_n}{t_n - \tau_u''}\right) : \left(1 + \frac{t_n^* - t_n}{t_n - \tau_n'}\right).$$

This yields the estimate

$$(22.4) \qquad |\mu_n - 1| \leq \frac{2\varepsilon_n}{l - \varepsilon_n}.$$

In place of (22.3) I can write

$$(22.5) \qquad \frac{1}{t^* - \tau_n'} = \lambda_n + \mu_n \frac{1}{t - \tau_n'} \qquad (\lambda_n \text{ constant});$$

for every linear substitution with the fixed point τ_n' must be of this form. Also if the transformation T_n is parabolic, I can put it in the form (22.5); in this case $\mu_n = 1$ and (22.4) certainly holds. We get the limit equation

$$\lim_{n=\infty} \mu_n = 1.$$

If again we set $t = t_n$ and $t^* = t_n^*$, now in (22.5), we find

$$\lambda_n = \frac{t_n - t_n^*}{(t_n - \tau_n')(t_n^* - \tau_n')} - \frac{\mu_n - 1}{t_n - \tau_n'},$$

$$|\lambda_n| \leq \frac{\varepsilon_n}{l^2} + \frac{|\mu_n - 1|}{l}, \qquad \lim_{n=\infty} \lambda_n = 0.$$

Finally, if we bring (22.5) into its natural form

$$t^* = \frac{\alpha_n t + \beta_n}{\gamma_n t + \delta_n},$$

we get the values

$$\alpha_n = 1 + \lambda_n \tau_n', \qquad \beta_n = \tau_n'(\mu_n - 1 - \lambda_n \tau_n'),$$
$$\gamma_n = \lambda_n, \qquad \delta_n = \mu_n - \lambda_n \tau_n',$$

which differ from the corresponding coefficients in the identity substitution

$$\begin{pmatrix} 1 & 0 \\ 0 & 1 \end{pmatrix}$$

by less than

$$|\mu_n - 1| + |\lambda_n|.$$

This completes the proof of the lemma.

Now let Γ^* be an n-E group of motions, *without infinitesimal transformations,* and let τ be a fixed point in the n-E plane of some rotation belonging to Γ^*. By S_τ I denote all rotations in Γ^* which have the fixed point τ; they form a group in themselves. If I normalize the angles of rotation ϕ of the S_τ by the condition $0 \leq \phi < 1$, then, because of the lack of infinitesimal transformations, there exists one of the S_τ, say S_τ^0, with the *smallest* angle of rotation $\phi_0 > 0$. The angle of rotation of every S_τ is an integral multiple of ϕ_0; for otherwise, by appropriate choice of the integer k, one could generate a transformation $S_\tau (S_\tau^0)^{-k}$ with a smaller angle of rotation than S_τ^0. Thus S_τ^0 is a "primitive" transformation in the group of the S_τ; all the others are obtained as powers of S_τ^0. If one determines the integer h by the condition

$$h\phi_0 \leq 1 < (h+1)\phi_0,$$

then clearly $(S_\tau^0)^{h+1}$ has a smaller angle of rotation, $(h + 1)\phi_0 - 1$, than S_τ^0, unless $h\phi_0 = 1$. Therefore $\phi_0 = 1/h$ must hold, and h is the *order* of the finite cyclic group (S_τ) which consists of the rotations

$$(S_\tau^0)^1, (S_\tau^0)^2, (S_\tau^0)^3, \ldots, (S_\tau^0)^h = 1.$$

The fixed points of the rotations in Γ^* can have no limit point in the finite. If τ ($|\tau| < 1$) were such a limit point and if

$$S_n\left(\text{angle of rotation} = \frac{1}{h_n}\right) \qquad (n = 1, 2, 3, \ldots)$$

were a sequence of primitive rotations in Γ^*, with distinct fixed points τ_n which converged to τ, then there would be two a priori possibilities.

(1) The orders h_n are bounded for all n. Then there would be infinitely many of the S_n which have the *same* order h; let them be the transformations $S_1', S_2', S_3', \ldots$. Then $S_n'(S_{n+1}')^{-1}$ is infinitesimal for infinitely large n. Therefore this case is ruled out.

(2) The orders h_n are not bounded. Then one can choose a subsequence S_n' of the S_n such that the associated orders tend to ∞ and the angles of rotation tend to zero. But that also would contradict the absence of infinitesimal operations in Γ^*.

Having settled this, we can now show that a system of equivalent points, under Γ^*, can have no limit point in the finite. Suppose that t_0 ($|t_0| < 1$) were the limit of a sequence of equivalent points t_n ($n = 1, 2, 3, \ldots$):

$$\lim_{n=\infty} t_n = t_0.$$

Let S_n be the motion of Γ^* which carries t_n into t_{n+1}. I pick a neighborhood $\mathfrak{U}_0$ of t_0 which, with the possible exception of t_0 itself, contains no fixed point of a rotation in Γ^*. By the lemma, the fixed point τ_n' ($|\tau_n'| \leq 1$) of S_n can lie outside of $\mathfrak{U}_0$ for only finitely many n. Therefore we must have $\tau_n' = t_0$ for $n \geq n_0$; it is no restriction to assume that $n_0 = 1$. Then all t_n result from t_1 by rotations about t_0, which are contained in Γ^*; but by such rotations I can carry t_1 into only finitely many different positions. Thus the possibility of a limit point is contradicted, and the proof of the theorem stated at the beginning of § 22 is complete.

With this preparation out of the way, we arrive at the real object of this final section of our exposition. We are concerned with the one-to-one conformal maps of a Riemann surface $\mathfrak{F}$ onto itself. These maps form a group; the significance of this group for the theory of the Riemann surface $\mathfrak{F}$ is obviously analogous to that, for example, of the group of motions for metric geometry. Therefore when we formulate the following theorem, due in essence to Klein, we are stating a fact of fundamental importance.

The group of conformal maps of a Riemann surface onto itself is always discontinuous, except for the following seven exceptional cases: $\mathfrak{F}$ = complete sphere, simply or doubly punched sphere, a disc on the sphere [literally, "skull cap on the sphere"], a punched disc (i.e., a disc without its center), a zone on the sphere (between two circles of latitude[51]), a closed Riemann surface of genus one.

Proof.[52] Let $\tilde{\mathfrak{F}}$ denote the universal covering surface over $\mathfrak{F}$, let t be the cut-circle uniformizing variable which maps $\tilde{\mathfrak{F}}$ conformally onto the open unit disc, and let Γ be the group of n-E motions belonging to the Riemann surface $\mathfrak{F}$. The cases in which the t-sphere or the infinite t-plane, instead of the unit disc, arises may be ignored; our theorem is obviously invalid in those cases (exceptional cases 1, 2, 3, 7). Let C be a conformal map of $\mathfrak{F}$ onto itself which carries the point $\mathfrak{p}_0$ into $\mathfrak{p}_0'$. Let $\tilde{\mathfrak{p}}_0$ be a point of $\tilde{\mathfrak{F}}$ over $\mathfrak{p}_0$, $\tilde{\mathfrak{p}}_0'$ a point of $\tilde{\mathfrak{F}}$ over $\mathfrak{p}_0'$. Because of the simple connectivity of $\tilde{\mathfrak{F}}$, there exists a single topological map $\tilde{C}$ of $\tilde{\mathfrak{F}}$ onto itself such that:

(1) $\tilde{\mathfrak{p}}_0$ goes into $\tilde{\mathfrak{p}}_0'$, and

(2) every point $\tilde{\mathfrak{p}}$ on $\tilde{\mathfrak{F}}$ goes into a point $\widetilde{\mathfrak{p}C}$, whose trace point $\mathfrak{p}C$ is the image under C of the trace $\mathfrak{p}$ of $\tilde{\mathfrak{p}}$.

---

[51]) These constitute, according to the breadth of the zone, infinitely many essentially different Riemann surfaces.

[52]) This is Poincaré's proof: Acta Math. 7 (1885) 16–19. To be sure, Poincaré considers only closed Riemann surfaces; but his proof, word for word, remains valid for open surfaces.

$\mathfrak{C}$ is also conformal, and therefore appears in the t-plane as a linear transformation T, which preserves the open unit disc; also T carries any system of equivalent points under Γ into a system of the same sort. This last fact is expressed by the equation

$$T\Gamma T^{-1} = \Gamma,$$

which states that T and Γ commute; or, for every substitution S in Γ, the "transform" TST^{-1} also belongs to Γ. Conversely, it is also clear that every linear substitution T that commutes with Γ, and which carries $|t| < 1$ onto itself, furnishes a conformal map of $\mathfrak{F}_\Gamma$ onto itself. Obviously, the set of motions of the n-E plane which commute with Γ form a *group*, Γ_c, which contains Γ as a subgroup. The discontinuity of this group Γ_c is to be proved; by the theorem at the beginning of this section, it suffices to demonstrate the absence of infinitesimal operations in Γ_c.

If S and T are any two linear transformations preserving the open unit disc, then the two points which T throws into the fixed points of S are the fixed points of the transformation

$$TST^{-1} = S'.$$

If S' has the *same* fixed points as S, then either T must carry each of the fixed points of S into itself (that is, if S is not parabolic, T must have the same fixed points as S; if S is parabolic, at least one of the fixed points of T must coincide with the fixed point of S); or T must interchange the two fixed points σ' and σ'' of the (nonparabolic) transformation S, that is, T must carry σ' into σ'' and σ'' into σ'. As is clear from (22.2), a parabolic transformation cannot accomplish this last; a nonparabolic T (22.1) can accomplish it only if $\mu^2 = 1$, $\mu = -1$, and T is an elliptic transformation with rotation angle $\frac{1}{2}$. Thus it is possible for S and T to commute

$$TS = ST, \qquad TST^{-1} = S,$$

only if

I. both transformations S and T are nonparabolic and have the same fixed points, or

II. both transformations S and T are parabolic and have the same fixed point, or

III. S and T are elliptic transformations with angles of rotation $= \frac{1}{2}$, or

IV. one of the two transformations, S and T, is the identity.

If Γ consists of only the identity, then $\mathfrak{F}$ is conformally equivalent to the open unit disc, or a disc on the sphere (exceptional case 4).

Now let S be an arbitrary substitution in Γ, distinct from the identity. Assume that, contrary to our claim, there is a sequence of operations $T_n \neq 1$

in Γ_c, such that T_n converges to the identity. Then both $T_n S T_n^{-1}$ and $T_n S T_n^{-1} S^{-1}$ would be elements of Γ. Along with T_n, the last of these operations converges to the identity as n tends to infinity. But since Γ contains no infinitesimal transformations, it must be that

$$T_n S T_n^{-1} S^{-1} = 1$$

holds for $n \geq n_0$; that is, S and T_n commute. We have listed four cases in which this is possible, but since S is not elliptic, only cases I and II remain in question. If we apply this result to all possible operations $S \neq 1$ of the group Γ, we find that only two possibilities remain open:

I. All operations in Γ are hyperbolic and have the same two fixed points, which one may assume are $+i$ and $-i$.

II. All operations in Γ are parabolic and have the same fixed point, which we may take to be $+i$.

(Case I). The map

$$z = \lg \frac{i - t}{i + t}$$

carries the open unit disc in the t-plane onto the parallel strip $-\pi/2 < y < +\pi/2$ in the $z = (x + iy)$-plane; Γ goes into a group of translations parallel to the x-axis; that is, Γ takes the form

$$(22.6) \qquad z^* = z + n\frac{2\pi}{a} \qquad \begin{pmatrix} a \text{ a positive constant;} \\ n = 0, \pm 1, \pm 2, \ldots \end{pmatrix}.$$

Then

$$w = e^{aiz}$$

maps $\mathfrak{F}_\Gamma$ one-to-one onto the annulus

$$e^{-\frac{1}{2}a\pi} < |w| < e^{\frac{1}{2}a\pi},$$

and by stereographic projection this may be turned into a zone on the sphere with the equator for its mid-line (exceptional case 6).

(Case II.) The map

$$z = \frac{2}{t - i} - i \qquad (z = x + iy)$$

carries the open unit disc in the t-plane into the upper half-plane $y > 0$, and Γ into a group of translations parallel to the x-axis. So again Γ has the form

(22.6) and $w = e^{aiz}$ carries $\mathfrak{F}_\Gamma$ onto the punched unit disc in the w-plane:

$$0 < |w| < 1 \qquad \text{(exceptional case 5).}$$

With this our principal theorem is completely proved; at the same time we have had the opportunity to observe the applicability of the normal form $\mathfrak{F}_\Gamma$ of the Riemann surface through an important example. It is trivial that in each of the seven excluded cases there is a continuous group of conformal maps of the surface onto itself. It is also easy to describe these groups completely.

A closed Riemann surface of genus $p > 1$ *admits only a finite number of conformal maps onto itself.*[53] For a system of equivalent points under this group can have no limit point on the closed surface, and therefore contains only a finite number of points.

---

[53]) For this more special theorem, first stated by H. A. Schwarz, there exist other more algebraic proofs. One of Weierstrass (1875), which was first published in 1895 (Werke, Vol. II, pp. 235–244); one of M. Noether (Math. Ann., **20**, pp. 59–62 and **21**, pp. 138–140; 1882); and one of A. Hurwitz, Math. Ann., **41** (1893) 403–411.

INDEX

Mathematics

FUNCTIONAL ANALYSIS (Second Corrected Edition), George Bachman and Lawrence Narici. Excellent treatment of subject geared toward students with background in linear algebra, advanced calculus, physics and engineering. Text covers introduction to inner-product spaces, normed, metric spaces, and topological spaces; complete orthonormal sets, the Hahn-Banach Theorem and its consequences, and many other related subjects. 1966 ed. 544pp. 6⅛ x 9¼. 0-486-40251-7

DIFFERENTIAL MANIFOLDS, Antoni A. Kosinski. Introductory text for advanced undergraduates and graduate students presents systematic study of the topological structure of smooth manifolds, starting with elements of theory and concluding with method of surgery. 1993 edition. 288pp. 5⅜ x 8½. 0-486-46244-7

VECTOR AND TENSOR ANALYSIS WITH APPLICATIONS, A. I. Borisenko and I. E. Tarapov. Concise introduction. Worked-out problems, solutions, exercises. 257pp. 5⅝ x 8¼. 0-486-63833-2

AN INTRODUCTION TO ORDINARY DIFFERENTIAL EQUATIONS, Earl A. Coddington. A thorough and systematic first course in elementary differential equations for undergraduates in mathematics and science, with many exercises and problems (with answers). Index. 304pp. 5⅜ x 8½. 0-486-65942-9

FOURIER SERIES AND ORTHOGONAL FUNCTIONS, Harry F. Davis. An incisive text combining theory and practical example to introduce Fourier series, orthogonal functions and applications of the Fourier method to boundary-value problems. 570 exercises. Answers and notes. 416pp. 5⅜ x 8½. 0-486-65973-9

COMPUTABILITY AND UNSOLVABILITY, Martin Davis. Classic graduate-level introduction to theory of computability, usually referred to as theory of recurrent functions. New preface and appendix. 288pp. 5⅜ x 8½. 0-486-61471-9

AN INTRODUCTION TO MATHEMATICAL ANALYSIS, Robert A. Rankin. Dealing chiefly with functions of a single real variable, this text by a distinguished educator introduces limits, continuity, differentiability, integration, convergence of infinite series, double series, and infinite products. 1963 edition. 624pp. 5⅜ x 8½. 0-486-46251-X

METHODS OF NUMERICAL INTEGRATION (SECOND EDITION), Philip J. Davis and Philip Rabinowitz. Requiring only a background in calculus, this text covers approximate integration over finite and infinite intervals, error analysis, approximate integration in two or more dimensions, and automatic integration. 1984 edition. 624pp. 5⅜ x 8½. 0-486-45339-1

INTRODUCTION TO LINEAR ALGEBRA AND DIFFERENTIAL EQUATIONS, John W. Dettman. Excellent text covers complex numbers, determinants, orthonormal bases, Laplace transforms, much more. Exercises with solutions. Undergraduate level. 416pp. 5⅜ x 8½. 0-486-65191-6

RIEMANN'S ZETA FUNCTION, H. M. Edwards. Superb, high-level study of landmark 1859 publication entitled "On the Number of Primes Less Than a Given Magnitude" traces developments in mathematical theory that it inspired. xiv+315pp. 5⅜ x 8½. 0-486-41740-9

CALCULUS OF VARIATIONS WITH APPLICATIONS, George M. Ewing. Applications-oriented introduction to variational theory develops insight and promotes understanding of specialized books, research papers. Suitable for advanced undergraduate/graduate students as primary, supplementary text. 352pp. $5\frac{3}{8}$ x $8\frac{1}{2}$.
0-486-64856-7

MATHEMATICIAN'S DELIGHT, W. W. Sawyer. "Recommended with confidence" by *The Times Literary Supplement*, this lively survey was written by a renowned teacher. It starts with arithmetic and algebra, gradually proceeding to trigonometry and calculus. 1943 edition. 240pp. $5\frac{3}{8}$ x $8\frac{1}{2}$. 0-486-46240-4

ADVANCED EUCLIDEAN GEOMETRY, Roger A. Johnson. This classic text explores the geometry of the triangle and the circle, concentrating on extensions of Euclidean theory, and examining in detail many relatively recent theorems. 1929 edition. 336pp. $5\frac{3}{8}$ x $8\frac{1}{2}$. 0-486-46237-4

COUNTEREXAMPLES IN ANALYSIS, Bernard R. Gelbaum and John M. H. Olmsted. These counterexamples deal mostly with the part of analysis known as "real variables." The first half covers the real number system, and the second half encompasses higher dimensions. 1962 edition. xxiv+198pp. $5\frac{3}{8}$ x $8\frac{1}{2}$. 0-486-42875-3

CATASTROPHE THEORY FOR SCIENTISTS AND ENGINEERS, Robert Gilmore. Advanced-level treatment describes mathematics of theory grounded in the work of Poincaré, R. Thom, other mathematicians. Also important applications to problems in mathematics, physics, chemistry and engineering. 1981 edition. References. 28 tables. 397 black-and-white illustrations. xvii + 666pp. $6\frac{1}{8}$ x $9\frac{1}{4}$. 0-486-67539-4

COMPLEX VARIABLES: Second Edition, Robert B. Ash and W. P. Novinger. Suitable for advanced undergraduates and graduate students, this newly revised treatment covers Cauchy theorem and its applications, analytic functions, and the prime number theorem. Numerous problems and solutions. 2004 edition. 224pp. $6\frac{1}{2}$ x $9\frac{1}{4}$. 0-486-46250-1

NUMERICAL METHODS FOR SCIENTISTS AND ENGINEERS, Richard Hamming. Classic text stresses frequency approach in coverage of algorithms, polynomial approximation, Fourier approximation, exponential approximation, other topics. Revised and enlarged 2nd edition. 721pp. $5\frac{3}{8}$ x $8\frac{1}{2}$. 0-486-65241-6

INTRODUCTION TO NUMERICAL ANALYSIS (2nd Edition), F. B. Hildebrand. Classic, fundamental treatment covers computation, approximation, interpolation, numerical differentiation and integration, other topics. 150 new problems. 669pp. $5\frac{3}{8}$ x $8\frac{1}{2}$.
0-486-65363-3

MARKOV PROCESSES AND POTENTIAL THEORY, Robert M. Blumental and Ronald K. Getoor. This graduate-level text explores the relationship between Markov processes and potential theory in terms of excessive functions, multiplicative functionals and subprocesses, additive functionals and their potentials, and dual processes. 1968 edition. 320pp. $5\frac{3}{8}$ x $8\frac{1}{2}$. 0-486-46263-3

ABSTRACT SETS AND FINITE ORDINALS: An Introduction to the Study of Set Theory, G. B. Keene. This text unites logical and philosophical aspects of set theory in a manner intelligible to mathematicians without training in formal logic and to logicians without a mathematical background. 1961 edition. 112pp. $5\frac{3}{8}$ x $8\frac{1}{2}$. 0-486-46249-8